A Treatise of Indian and Tropical Soils

D.K. Pal

A Treatise of Indian and Tropical Soils

 Springer

D.K. Pal
ICAR-NBSS&LUP
Nagpur, Maharashtra
India

ISBN 978-3-319-49438-8   ISBN 978-3-319-49439-5   (eBook)
DOI 10.1007/978-3-319-49439-5

Library of Congress Control Number: 2016959255

Printed on acid-free paper

This Springer imprint is published by Springer Nature
The registered company is Springer International Publishing AG
The registered company address is: Gewerbestrasse 11, 6330 Cham, Switzerland

I dedicate this treatise to my parents and parents-in-law

Preface

The main purpose of this treatise lies in applying the basic principles of pedology to soils of the tropical environments of the Indian subcontinent, with emphasis on ways to enhance crop productivity. Although much valuable work has been done throughout the tropics, it has been always difficult to manage these soils to sustain their productivity and it is more so when comprehensive knowledge on their formation remained incomplete for a long time. During the last few decades, research endeavour undertaken by the Indian pedologists and earth scientists of well-known institutions, especially by those in the ICAR-NBSS&LUP (Indian Council of Agricultural Research-National Bureau of Soil Survey and Land Use Planning), Department of Earth Sciences of the erstwhile University of Roorkee (presently known as Indian Institute of Technology, Roorkee) and Department of Geology, University of Delhi, has been commendable in terms of establishing an organic link between pedology, geomorphology, palaeopedology, mineralogy and edaphology of Indian tropical soils. This treatise thus makes a sincere attempt to showcase the research contributions on pedology, geomorphology, mineralogy, micromorphology and climate change collected from the published literature on three major soil types, i.e. shrink–swell soils, red ferruginous soils and soils of the Indo-Gangetic Plain(IGP) that occur in the tropical environments. The synthesis of literature attempts to provide insights into several aspects of five pedogenetically important soil orders like Alfisols, Mollisols, Ultisols, Vertisols and Inceptisols of tropical Indian environments. A special attention has been made to document the significance of minerals in soils and their overall influence in soil science in terms of pedology, palaeopedology, polygenesis and edaphology. Such knowledge base becomes critical when attempts are made to fill up the gap between food production and population growth.

A dire need for such a treatise has been felt amidst the myth on the formation of tropical soils in general and those in the Indian subcontinent in particular. Many such soils did experience the change of climate from humid to semi-arid environments in the Holocene period, and thus, their polygenetic history adds further challenges to soil/earth science researchers. In India, students of pedology (soil science) and pedogeomorphology generally come across extreme difficulties in relating to examples applying the principles of soil science from textbooks devoted almost exclusively to soils of temperate climate of erstwhile Soviet Union, Europe and the USA. Therefore, the format is arranged for a process-oriented treatise as a reference for pedologists, allied earth scientists and M.Sc. and Ph.D. students and also for land resource managers who are engaged in enhancing the productivity of such tropical soils in changing scenario of climate change in India and elsewhere.

The author will remain ever grateful to late Prof. S.K. Mukherjee and late Prof. B.B. Roy, who held the coveted position of Acharya P.C. Ray Professor of Agricultural Chemistry at the University of Calcutta, for drawing him to soil research that offered more than a lifetime of fascinating problems to unravel. The treatise is primarily based on significant research contributions made by the author's esteemed colleagues, namely Drs. T. Bhattacharyya, P. Chandran, S.K. Ray, C. Mandal, D.C. Nayak, Pramod Tiwari and late Dr. D.K. Mandal of ICAR-NBSS&LUP, Nagpur; Prof. Pankaj Srivastava of Geology Department, Delhi University; and many other colleagues at the ICAR-NBSS&LUP, Nagpur, and by several M.Sc. and Ph.D. students both at ICAR-NBSS&LUP and Department of Geology, Delhi University. Unstinted technical support and assistance received from Messrs. S.L. Durge, G.K. Kamble and L.M. Kharbikar helped the author enormously in bringing the task to a successful fruition.

The thoughts presented in this treatise have evolved over a period of years in discussion among the author and several prominent soil scientists. Special mention is due to Dr. D.R. Bhumbla, late Dr. J.S.P. Yadav and Dr. I.P. Abrol, former directors, Central Soil Salinity Research Institute, Karnal; late Dr. R.S. Murthy, late Dr. J.L. Sehgal and Dr. M. Velayutham, former directors; late Mr. J.C. Bhattacharjee, Dr. S.B. Deshpande, Dr. V.A.K. Sarma and Dr. A.R. Kalbande, former principal scientists of ICAR-NBSS&LUP, Nagpur; Dr. K.V. Raman, former president, Clay Minerals Society of India; former Prof. Kunal Ghosh, Calcutta University; Dr. R.K. Gupta (CIMMYT, New Delhi);late Dr. K.L. Sahrawat (ICRISAT, Telangana) and Dr. S.P. Wani (ICRISAT, Telangana) of CGIAR; Dr. D.L.N. Rao, principal scientist, ICAR-IISS, Bhopal, India; late Dr. H. Eswaran, SCS, USDA; and late Prof. Don H. Yaalon, Hebrew University, Israel.

The author duly acknowledged the sources of the diagrams and tables that have been adapted mostly from his publications.

To my wife, Banani, my daughters Deedhiti and Deepanwita, my brother-in-law Dhrubajyoti and my son-in-laws Jai and Nachiket, I am grateful for their patience, understanding, encouragement and above all unstinted moral support.

Nagpur, India   D.K. Pal

Contents

1 **Indian Tropical Soils: An Overview**.   1
   References.   5

2 **Cracking Clay Soils (Vertisols): Pedology, Mineralogy
   and Taxonomy**.   9
   2.1   Introduction   10
   2.2   Factors of Vertisol Formation: A Revisit   12
         2.2.1   Parent Material   12
         2.2.2   Climate.   13
         2.2.3   Topography.   14
         2.2.4   Vegetation.   19
         2.2.5   Time   19
   2.3   Smectite Clay Minerals in Vertisols: Recent Initiatives
         for Their Proper Characterization   20
         2.3.1   Characteristics of Smectite   20
         2.3.2   Seat of Charge in and Layer Charge of Fine Clay
                 Smectites.   22
         2.3.3   Hydroxy-Interlayered Smectites (HIS) and Determination
                 of Layer Charge   25
   2.4   Genesis of Smectites in Vertisols: A Revised Status.   25
   2.5   Recent Advances in Pedogenic Processes in Vertisols   28
         2.5.1   Clay Illuviation in Vertisols: An Example of
                 Proanisotropism.   28
         2.5.2   Factors of Clay Illuviation in Vertisols   30
         2.5.3   Relative Rapidity of Clay Illuviation, Pedoturbation
                 and Slickenside Formation   30
         2.5.4   Development of Microstructures and Vertical Cracks
                 in Vertisols as Controlled by Smectite Swelling   34
   2.6   Evolutionary Pathways of Vertisol Formation.   35
   References.   37

3   Red Ferruginous Soils: Pedology, Mineralogy and Taxonomy......   43
  3.1   Introduction   44
  3.2   US Soil Taxonomic Classes and Factors in the Formation of Red
        Ferruginous (RF) Soils...................................   48
  3.3   Clay Illuviation, Clay Enriched B Horizons and Taxonomic
        Rationale of RF Soils of Humid Tropical (HT) Climate.........   49
  3.4   RF Soils (Alfisols) in Semi-arid Tropics (SAT): Record
        of Climate Change and Polypedogenesis in the Holocene Period ...   51
  3.5   Acidity and Its Nature and Charge Characteristics in RF Soils
        of HT Climate: The Present Status.........................   56
  3.6   Clay Mineralogy of RF Soils of the HT Climate: Recent
        Advances   61
        3.6.1   Genesis of Gibbsite at the Expense of Kaolinite:
                Dissolution of a Myth.............................   61
        3.6.2   Mixed Mineralogy Class for RF Soils of HT Climate:
                A Rational Proposal   64
  3.7   Ultisols of Indian Tropical Humid Climate in Context
        with Doubtful Existence of Oxisols   65
  References..   66

**4   Soils of the Indo-Gangetic Alluvial Plains: Historical Perspective,
    Soil-Geomorphology and Pedology in Response Climate Change
    and Neotectonics**   71
  4.1   Introduction   72
  4.2   Early Studies on Soils   74
  4.3   Soil-Geomorphology of the Indo-Gangetic Plains (IGP).........   74
  4.4   Soil Ages of the IGP   76
  4.5   Pedogenic Response of the IGP Soils to Holocene Climatic
        Fluctuations   77
  4.6   Pedogenic Response of the IGP Soils to Neotectonic Events   79
  4.7   Soils of the IGP: Bio-climates, Pedogenic Processes, Mineralogy
        and Taxonomic Classes   81
        4.7.1   Development of Soils in Different Bio-climates   81
        4.7.2   Pedogenic Processes, Polygenesis and Taxonomic
                Classes of the IGP Soils   82
        4.7.3   Clay Minerals and Mineralogy Class of the IGP Soils   84
  References..   87

5   Conceptual Models on Tropical Soil Formation   91
  5.1   Introduction   91
  5.2   Vertisols Spatially Associated with Alfisols and Mollisols
        on Zeolitic Deccan Basalts................................   92
  5.3   Acidic Alfisols, Mollisols and Ultisols on Gneissic
        and Sedimentary Rock Systems   92
  5.4   Critique on the Validation of the Conceptual Models..........   93
  References..   94

6 Land and Soil Degradation and Remedial Measures 97
  6.1 Introduction 97
  6.2 Natural Chemical Degradation of Soils in the Indian SAT,
     and Remedial Measures 98
  6.3 Definition, Processes and Factors of Soil Degradation 99
  6.4 Natural Chemical Soil Degradation: Neotectonic-Climate Linked
     and Mineral Induced.................................... 100
  6.5 Chemical Soil Degradation: Regressive Pedogenic Processes
     in the SAT 101
  6.6 Clay Mineral (Palygorskite) Induced Natural Chemical Soil
     Degradation 103
  6.7 Micro-topography: A Unique Factor of Natural Soil Degradation
     in SAT 104
  6.8 Indices of Soil Degradation 106
  6.9 Rehabilitation of Degraded Soils 107
    6.9.1 Difficulties in Identification and Reclamation of Sodic
        Soils 107
    6.9.2 An Alternative Management to Reclaim Sodic Soils. 107
    6.9.3 Linking Calcium Carbonates to Resilience of the IGP
        Sodic Soils 108
    6.9.4 Calcium Carbonates as Soil Modifier in Ensuring Soil
        Sustainability of SAT Soils........................ 109
  6.10 Nature and Extent of Degradation in RF Soils: A Critique. 109
  References. 112

**7 Clay and Other Minerals in Soils and Sediments as Evidence
of Climate Change.** 115
  7.1 Introduction 115
  7.2 Di- and Tri-Octahedral Smectite as Evidence for Paleoclimatic
     Changes 117
  7.3 Red and Black Soils in Semi-arid Climatic Environments 118
  7.4 Clay Minerals in Soils of the Indo-Gangetic Plains (IGP) 118
  7.5 Vertisols, Carbonate Minerals and Climate Change 119
  7.6 Implications for Climate Change from Clay Minerals Record
     in Drill Cores of the Ganga Plains 122
  References. 123

8 Linking Minerals to Selected Soil Bulk Properties 127
  8.1 Introduction 128
  8.2 Clay and Other Minerals in Adsorption and Desorption of Major
     Nutrients. 128
    8.2.1 Organic and Inorganic Carbon 128
    8.2.2 Nitrogen 132
    8.2.3 Phosphorus 134
    8.2.4 Potassium 137
  References. 148

**9   Importance of Pedology of Indian Tropical Soils in Their
     Edaphology** 153
     9.1   Introduction 153
     9.2   Impact of Spatially Associated Non-sodic (Aridic Haplusterts)
           and Sodic (Sodic Haplusterts) Vertisols on Crop Performance 154
     9.3   Linear Distance of Cyclic Horizons in Vertisols and Its
           Relevance to Agronomic Practices 155
     9.4   Smectite as the First Weathered Product of the Deccan Basalt
           and Its Contribution in Growing Vegetation on Weathered Basalt
           and in Very Shallow Cracking Clay Soil 157
     9.5   Sustainability of Rice Production in Zeolitic Vertisols 159
     9.6   Holocene Climate Change, Natural Soil Degradation, Modified
           Vertisols and Their Evaluation 160
     9.7   Soil Modifiers (Zeolites, Gypsum and $CaCO_3$) in Mitigating
           the Adverse Holocene Climate Change and Making Sodic
           Vertisols Resilient. 162
           9.7.1 Ca-Zeolites 162
           9.7.2 Gypsum. 164
           9.7.3 $CaCO_3$ 165
     9.8   Acidity, Al Toxicity and Lime Requirement in RF Soils of HT
           Climates: A Critique. 167
     9.9   Present Soil Health Status Due to Anthropogenic Activities
           in IGP Soils 168
     References. 170

10 Summary and Concluding Remarks 175

About the Author

Dr. D.K. Pal graduated in 1968 with honours in Chemistry, obtained M.Sc. (Ag) degree in agricultural chemistry with specialization in soil science in 1970 and earned his Ph.D. degree in agricultural chemistry in 1976 from the Calcutta University. He worked as a DAAD postdoctoral fellow at the Institute of Soil Science, University of Hannover, West Germany, during 1980–1981.

Research activities of Dr. Pal have spanned more than three and a half decades and focused on the alluvial (Indo-Gangetic Plain, IGP), red ferruginous and shrink–swell soils of the tropical environments of India. His work has expanded the basic knowledge in pedology, palaeopedology, soil taxonomy, soil mineralogy, soil micromorphology and edaphology. He also created an internationally recognized school of thought on the development and management of the Indian tropical soils as evidenced by significant publications in several leading international journals in soil, clay and earth sciences inviting critical appreciation by the peers.

Dr. Pal has built an excellent research team in mineralogy, micromorphology, pedology and palaeopedology in the country. Under his guidance, the team members have carved out a niche for themselves in soil research at national and international level. He has mentored several M.Sc. and Ph.D. students of land resource management (LRM) of the Indian Council of Agricultural Research-National Bureau of Soil Survey and Land Use Planning (ICAR-NBSS&LUP) under the academic programme at Dr. Panjab Rao Deshmukh Krishi Vidyapeeth (Dr. P D K V), Akola.

He has delivered numerous prestigious invited lectures at national and international meets. In addition, he has served as a reviewer for many journals of national and international repute and contributed reviews and book chapters for national and international publishers. He has been coeditor of books, proceedings and journals. He is the life member of many professional national societies in soil and earth science. He has also been conferred with awards (The Platinum Jubilee Commemoration Award of the Indian Society of Soil Science, New Delhi, for the year 2012, ICAR Award, Outstanding Interdisciplinary Team Research in Agriculture and Allied Sciences, Biennium 2005–2006, and the 12th International Congress Commemoration Award, Indian Society of Soil Science, 1997) and fellowships (West Bengal Academy of Science and Technology, 2014, National Academy of Agricultural Sciences, New Delhi, 2010, Indian Society of Soil Science, New Delhi, 2001, and Maharashtra Academy of Sciences, Pune, 1996).

He worked as the Principal Scientist and Head, Division of Soil Resource Studies, ICAR-NBSS&LUP, Nagpur, and as a visiting scientist at the International Crops Research Institute for the Semi-Arid Tropics (ICRISAT), Telangana.

Chapter 1
Indian Tropical Soils: An Overview

Abstract The Indian subcontinent, which collided with the Asian mainland during the Eocene period, is a very old mass and has not been under water since the Carboniferous period. A girdle of high mountains, snow fields, glaciers and thick forests in the north, seas washing lengthy coasts in the Peninsula, a variety of geological formations, diversified climate, topography and relief have given rise to varied physiographic features. Temperature varies from arctic cold to equatorial hot. Such varied natural environments have resulted in a great variety of soils in India compared to any other country of similar size in the world. Many however think of tropical soils as the deep red and highly weathered soils, and are often thought are either agriculturally poor or virtually useless. The major soils of India are Vertisols, Mollisols, Alfisols, Ultisols, Aridisols, Inceptisols and Entisols. Although soils of India occur in 5 bio-climatic systems, but only a few soil orders are spread in more than one bio climate. Vertisols belong to arid hot, semi-arid, sub-humid and humid to per-humid climatic environments. Mollisols belong to sub-humid and also humid to per-humid climates. Alfisols belong to semi-arid, sub-humid and also in humid to per-humid climates, whereas Ultisols belong to only humid to per-humid climates. Both Entisols and Inceptisols belong to all the 5 categories of bio-climatic zones of India, and Aridisols belong mainly to arid climatic environments. This baseline information indicates that except for the Ultisols and Aridisols, the rest 5 soil orders exist in more than one bio-climatic zones of India. The absence of Oxisols and a small area under Ultisols, suggest that soil diversity in the geographic tropics in India, is as large as in the temperate zone. These soils are not confined to a single production system and generally maintain a positive organic carbon balance. Thus they contribute substantially to India's growing self-sufficiency in food production and food stocks. Therefore, any generalizations about tropical soils are unlikely to have wider applicability in the Indian subcontinent. The genesis of Ultisols alongside acidic Alfisols and Mollisols for the millions of years in both zeolitic and non zeolitic parent materials in Indian humid tropical (HT) climatic environments indicates how the parent material composition influences the formation of Alfisols, Mollisols and Ultisols in weathering environments of HT climate; and also how the relict Alfisols of semi-arid tropical (SAT) environments are polygenetic. The critical evaluation of the nature and distribution of naturally occurring clay minerals,

© Springer International Publishing AG 2017
D.K. Pal, *A Treatise of Indian and Tropical Soils*,
DOI 10.1007/978-3-319-49439-5_1

calcium carbonates, gypsum, gibbsite and zeolites can yield valuable and important information to comprehend the complex factors involved in the pedogenesis of soils formed in the present and past climates. Thus, the conventional management protocols to improve and sustain their productivity need to be revised in the light of new knowledge gained in recent years. Global distribution of tropical soils and the recent advances in knowledge by researching on them (Entisols, Inceptisols, Mollisols, Alfisols, Vertisols and Ultisols) in the Indian sub-continent indicates that some of the agricultural management practices developed in this part of the tropical world for enhancing crop productivity and maintaining soil health, might also be adoptable to similar soils elsewhere. In the following chapters from 2 to 9, arguments are presented in terms readily understood by all stake holders of tropical soils and with both scientific and economic rigor so that they are not easily refuted.

Keywords Indian tropical soils · Advances in pedology · Climate change · Edaphology · Soil modifiers

Tropical soils are those soils that occur in geographic tropics (that part of the world located between 23.5° north and south of the Equator), meaning simply, the region of the earth between the Tropic of Cancer and the Tropic of Capricorn. This region is also known as Torrid Zone. The Indian subcontinent, which collided with the Asian mainland during the Eocene period, is a very old mass and has not been under water since the Carboniferous period. A girdle of high mountains, snow fields, glaciers and thick forests in the north, seas washing lengthy coasts in the Peninsula, a variety of geological formations, diversified climate, topography and relief have given rise to varied physiographic features. Temperature varies from arctic cold to equatorial hot; rainfall from barely a few centimetres in the arid parts, to per-humid with world's maximum rainfall of several hundred centimetres per annum in some other parts. These conditions provide for a landscape of high plateaus, stumpy relic hills, and shallow open valleys, rolling uplands, fertile plains, swampy low lands and dreary barren deserts. Such varied natural environments have resulted in a great variety of soils in India compared to any other country of similar size in the world (Bhattacharyya et al. 2013).

   Major part of the land area in India is however, in the region lying between the Tropic of Cancer and Tropic of Capricorn, and the soils therein are termed "tropical soils". Many however think of tropical soils as the soils of the hot and humid tropics only, exemplified by deep red and highly weathered soils and are often thought improperly they are either agriculturally poor or virtually useless (Sanchez 1976; Eswaran et al. 1992). India has 5 distinct bioclimatic systems (Bhattacharjee et al. 1982) with varying MAR; and they are arid cold and hot (MAR < 550 mm), semi-arid (MAR 550–1000 mm), sub-humid (MAR 1000–1500 mm), humid to per-humid (MAR 1200–3200 mm) and coastal (MAR 900–3000 mm). The major soils of India are Vertisols, Mollisols, Alfisols, Ultisols, Aridisols, Inceptisols and Entisols covering 8.1, 0.5, 12.8, 2.6, 4.1, 39.4 and 23.9%, respectively of the total geographical area (TGA) of the country (Bhattacharyya et al. 2009). However, the

Andisols in the humid tropical (HT) Nilgiri Hills, southern India, are non-allophanic Andisols derived from a special kind of non-volcanic material consisting of low-activity clay (LAC) residuum, rich in Al and Fe oxides, and deserve a special mention. LAC soils and the most common volcanic Andisols are often considered to be representative of advanced and juvenile stages of soil formation, respectively. The occurrence of these soils was unexpected. Secondary oxides, inherited from a previous cycle of soil genesis, appear to play the same role as volcanic glasses do in most Andisols (Caner et al. 2000). Andisols are not however, considered as one of the major soils of India because they are not mappable in 1:250,000.

Although soils of India occur in 5 bio-climatic systems, but only a few soil orders are spread in more than one bio climate. Vertisols belong to arid hot, semi-arid, sub-humid and humid to per-humid climatic environments (Bhattacharyya et al. 2005; Pal et al. 2009a). Mollisols belong to sub-humid and also humid to per-humid climates (Bhattacharyya et al. 2006). Alfisols belong to semi-arid, sub-humid and also in humid to per-humid climates (Pal et al. 1989, 1994, 2003; Bhattacharyya et al. 1993, 1999), whereas Ultisols belong to only humid to per-humid climates (Bhattacharyya et al. 2000; Chandran et al. 2005). Both Entisols and Inceptisols belong to all the 5 categories of bio-climatic zones of India, and Aridisols belong mainly to arid climatic environments (Bhattacharyya et al. 2008). This baseline information indicates that except for the Ultisols and Aridisols, the rest 5 soil orders exist in more than one bio-climatic zones of India. The absence of Oxisols and the Ultisols, occupying only 2.56% of total geographical area of the country, suggest that soil diversity in the geographic tropics in general and in India in particular, is at least as large as in the temperate zone (Eswaran et al. 1992; Sanchez and Logan 1992). These soils are not confined to a single production system and generally maintain a positive organic carbon (OC) balance without adding significantly to greenhouse gas emissions (Pal et al. 2015). Thus they contribute substantially to India's growing self-sufficiency in food production and food stocks (Pal et al. 2015; Bhattacharyya et al. 2014). Therefore, India can be called a land of paradoxes because of the large variety of soils and any generalizations about tropical soils are unlikely to have wider applicability in the Indian subcontinent (Pal et al. 2012a, 2014, 2015).

Mohr et al. (1972) stated that the parent rock seems to influence soil formation in such a way that similar soils are formed under quite different climatic conditions. However, an extensive pedogenetic study of Vertisols in an Indian climosequence (from arid hot to humid bio climates) expands the basic understanding of Vertisol evolution from Typic Haplusterts to Udic/Aridic/Sodic Haplusterts and Sodic Calciusterts (Pal et al. 2009a, 2012a). The genesis of Vertisols and soils with vertic character in the extra-peninsular region like in the micaceous alluvium of the IGP area needs to be reconciled through the role of Cratonic flux (Tandon et al. 2008; Pal et al. 2012b). While the formation of Vertisols in humid tropical (HT) climate is possible in the alluvium of zeolitic Deccan basalt, the formation of sodic shrink-swell soils (Sodic Haplusterts and Sodic Calciusterts, Pal et al. 2009a), sodic IGP soils (Typic Natrustalfs/Natraqualfs, Pal et al. 2003), and sodic red ferruginous (RF) soils (Typic Natrustalfs, Chandran et al. 2013) in semi-arid tropical

(SAT) environments is caused by the tectonic-climate linked natural soil degradation process (Pal et al. 2009b). These Vertisols, RF and IGP soils may remain in equilibrium with their climatic environments until the climate changes further, after which another pedogenic threshold is reached.

It is also believed that differences in rock composition lead to the formation of different soils even if the climate is similar over the whole area of study. But, Chesworth (1973) opined that the effect of the composition of parent rock on the composition of resulting soil in HT climate is an inverse function of time, and given enough time the chemical effect of parent rock would be nullified. In contrast, the genesis of Ultisols alongside acidic Alfisols and Mollisols for the millions of years in both zeolitic and non zeolitic parent materials in Indian HT climatic environments indicates how the parent material composition influences the formation of Alfisols, Mollisols and Ultisols in weathering environments of HT climate; and also how the relict Alfisols of SAT environments are polygenetic (Pal et al. 2014). These are some important but diverse issues on understanding the soils, tropical soils in particular.

It is generally understood that coarse-grained rocks tend to weather faster than fine-grained ones, and basic rocks faster than acid ones. Soils formed from basic rocks, as compared to those from acid rocks, are usually more fertile. The distribution of different geological rock formations (Fig. 1.1a) vis-a-vis different soils of India (Fig. 1.1b) reflects the relation between rocks and soils thereby suggesting the effect of parent rock on soil formation. For example, the fine textured Deccan basalt has been responsible for the formation of black (shrink-swell) soils whereas coarse-grained metamorphic rocks have given rise to red ferruginous soils (Pal et al. 2000a). However, the occurrence of spatially associated red ferruginous soils (Alfisols) in areas dominated by black soils (Vertisols) and vice versa almost under same topographical situation reflects two contrasting chemical environments that were conducive for the formation of these two groups of soils on the same parent material presumably under similar climatic conditions. This enigmatic but interesting fact has been a topic of researches the world over (Mohr et al. 1972). Research attempts made in India (Pal 1988, 2008; Bhattacharyya et al. 1993) and elsewhere (Beckmann et al. 1974) suggest their formation through a progressive landscape reduction process.

It has been recently demonstrated that the critical evaluation of the nature and distribution of naturally occurring clay minerals, calcium carbonates, gypsum, gibbsite and zeolites resulting either from inheritance or from the weathering of parent material whether mineral or rock, can yield valuable and important information to comprehend the complex factors involved in the pedogenesis of soils formed in the present and past climates (Birkland 1977; Rengasamy et al. 1978; Murali et al. 1978; Singer 1980; Pal et al. 1989, 2000a, b, 2001, 2009a, b, 2012a, b, 2014; Bhattacharyya et al. 1993, 1999, 2000; Chandran et al. 2005; Jenkins 1985; Wilson 1985). These situations emphasize the fact that since soils are the product of chemical, biochemical and physical processes acting upon earth materials under various topographic and climatic conditions, they bear the signatures as much as do landforms, the climatic and geomorphic history of the region in which they are

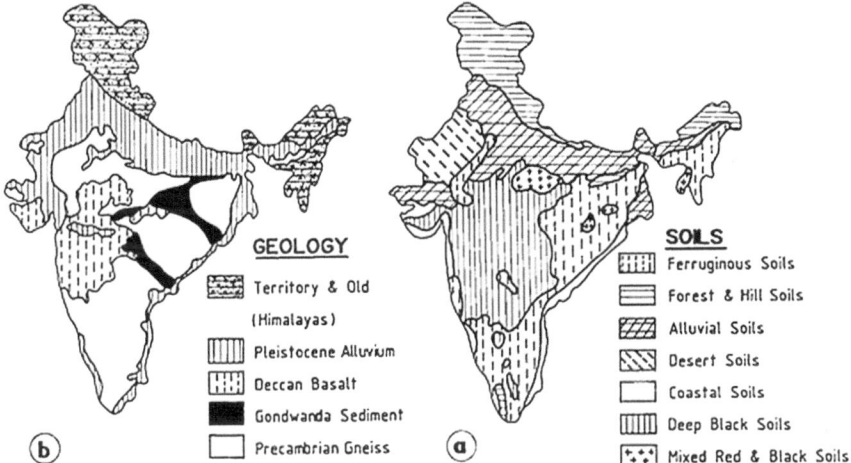

Fig. 1.1 Distribution of different kinds of soil (**a**) and the spread of geological formations (**b**) (Adapted from Pal et al. 2000a)

evolved (Thornbury 1969). Thus, the edaphological issues of soils of tropical environments in terms of following the conventional management protocols to improve and sustain their productivity need to be revised in the light of new knowledge gained in recent years (Pal et al. 2012a, c, 2014; Srivastava et al. 2015).

As the tropics comprise approximately 40% of the land surface of the earth, more than one-third of the soils of the world are tropical soils (Eswaran et al. 1992). Global distribution of these soils and the recent advances in knowledge by researching on them (Entisols, Inceptisols, Mollisols, Alfisols, Vertisols and Ultisols) in the Indian sub-continent indicates that some of the agricultural management practices developed in this part of the tropical world for enhancing crop productivity and maintaining soil health, for example, through carbon sequestration and other management interventions might also be adoptable to similar soils elsewhere. Arguments are presented in terms readily understood by all stake holders of tropical soils and with both scientific and economic rigor so that they are not easily refuted (Greenland 1991).

References

Beckmann GG, Thompson CH, Hubble GD (1974) Genesis of red and black soils on basalt on the Darling Downs, Queensland, Australia. J Soil Sci 25:265–280

Bhattacharjee JC, Roychaudhury C, Landey RJ, Pandey S (1982) Bioclimatic analysis of India. NBSSLUP Bulletin 7, National bureau of soil survey and land use planning (ICAR), Nagpur, India, p 21+ map

Bhattacharyya T, Pal DK, Deshpande SB (1993) Genesis and transformation of minerals in the formation of red (Alfisols) and black (Inceptisols and Vertisols) soils on Deccan Basalt in the Western Ghats, India. J Soil Sci 44:159–171

Bhattacharyya T, Pal DK, Srivastava P (1999) Role of zeolites in persistence of high altitude ferruginous Alfisols of the humid tropical Western Ghats, India. Geoderma 90:263–276

Bhattacharyya T, Pal DK, Srivastava P (2000) Formation of gibbsite in presence of 2:1 minerals: an example from Ultisols of northeast India. Clay Miner 35:827–840

Bhattacharyya T, Pal DK, Chandran P, Ray SK (2005) Land-use, clay mineral type and organic carbon content in two Mollisols–Alfisols–Vertisols catenary sequences of tropical India. Clay Res 24:105–122

Bhattacharyya T, Pal DK, Lal S, Chandran P, Ray SK (2006) Formation and persistence of Mollisols on zeolitic Deccan basalt of humid tropical India. Geoderma 146:609–620

Bhattacharyya T, Pal DK, Chandran P, Ray SK, Mandal C, Telpande B (2008) Soil carbon storage capacity as a tool to prioritise areas for carbon sequestration. Curr Sci 95:482–494

Bhattacharyya T, Sarkar D, Sehgal J, Velayutham M, Gajbhiye KS, Nagar AP, Nimkhedkar SS (2009) Soil taxonomic database of India and the states (1:250,000 scale). N B S S & L U P Publication 143, 266 pp

Bhattacharyya T, Pal DK, Mandal C, Chandran P, Ray SK, Sarkar D, Velmourougane K, Srivastava A, Sidhu GS, Singh RS, Sahoo AK, Dutta D, Nair KM, Srivastava R, Tiwary P, Nagar AP, Nimkhedkar SS (2013) Soils of India: historical perspective, classification and recent advances. Curr Sci 104:1308–1323

Bhattacharyya T, Chandran P, Ray SK, Mandal C, Tiwary P, Pal DK, Wani SP, Sahrawat KL (2014) Processes determining the sequestration and maintenance of carbon in soils: a synthesis of research from tropical India. Soil Horiz, Published 9 July 2014, pp 1–16. doi:10.2136/sh14-01-0001

Birkland PW (1977) Pedology, weathering, and geomorphological research. Oxford University Press, New York

Caner L, Bourgeon G, Toutain F, Herbillon AJ (2000) Characteristics of non-allophanic Andisols derived from low-activity clay regoliths in the Nilgiri Hills (Southern India). Eur J Soil Sci 51:553–563

Chandran P, Ray SK, Bhattacharyya T, Srivastava P, Krishnan P, Pal DK (2005) Lateritic Soils of Kerala, India: their mineralogy, genesis and taxonomy. Aust J Soil Res 43:839–852

Chandran P, Ray SK, Bhattacharyya T, Tiwari P, Sarkar D, Pal DK, Mandal C, Nimkar A, Maurya UK, Anantwar SG, Karthikeyan K, Dongare VT (2013) Calcareousness and subsoil sodicity in ferruginous Alfisols of southern India: an evidence of climate shift. Clay Res 32:114–126

Chesworth W (1973) The parent rock effect in the genesis of soil. Geoderma 10:215–225

Eswaran H, Kimble J, Cook T, Beinroth FH (1992) Soil diversity in the tropics: implications for agricultural development. In: Lal R, Sanchez PA (eds) Myths and science of soils of the tropics. SSSA special publication number 29. SSSA, Inc and ACA, Inc., Madison, pp 1–16

Greenland DJ (1991) The contributions of soil science to society—past, present, and future. Soil Sci 151:19–23

Jenkins DA (1985) Chemical and mineralogical composition in the identification of paleosols. In: Boardman J (ed) Soils and quaternary landscape evolution. Wiley, New York, pp 23–43

Mohr ECJ, Van Baren FA, Van Schuylenborgh J (1972) Tropical soils—a comprehensive study of their genesis. Mouton-Ichtiarbaru-Van Hoeve, The Hague

Murali V, Krishnamurti GSR, Sarma VAK (1978) Clay mineral distribution in two topo sequences of tropical soils of India. Geoderma 20:257–269

Pal DK (1988) On the formation of red and black soils in southern India. In: Hirekerur LR, Pal DK, Sehgal JL, Deshpande SB (eds) Transactions international workshop swell-shrink soils. Oxford & IBH, New Delhi, pp 81–82

Pal DK (2008) Soils and their mineral formation as tools in paleopedological and geomorphological studies. J Indian Soc Soil Sci 56:378–387

Pal DK, Deshpande SB, Venugopal KR, Kalbande AR (1989) Formation of di- and trioctahedral smectite as evidence for paleo-climatic changes in southern and central peninsular India. Geoderma 45:175–184

Pal DK, Kalbande AR, Deshpande SB, Sehgal JL (1994) Evidence of clay illuviation in sodic soils of Indo-Gangetic plain since the Holocene. Soil Sci 158:465–473

Pal DK, Bhattacharyya T, Deshpande SB, Sarma VAK, Velayutham M (2000a) Significance of minerals in soil environment of India, NBSS review series 1. NBSS&LUP, Nagpur 68p

Pal DK, Dasog GS, Vadivelu S, Ahuja RL, Bhattacharyya T (2000b) Secondary calcium carbonate in soils of arid and semi-arid regions of India. In: Lal R, Kimble JM, Eswaran H, Stewart BA (eds) Global climate change and pedogenic carbonates. Lewis Publishers, Boca Raton, USA, pp 149–185

Pal DK, Balpande SS, Srivastava P (2001) Polygenetic Vertisols of the Purna Valley of Central India. Catena 43:231–249

Pal DK, Srivastava P, Bhattacharyya T (2003) Clay illuviation in calcareous soils of the semi-arid part of the Indo-Gangetic Plains, India. Geoderma 115:177–192

Pal DK, Bhattacharyya T, Chandran P, Ray SK (2009a) Tectonics-climate-linked natural soil degradation and its impact in rainfed agriculture: Indian experience. In: Wani SP, Rockström J, Oweis T (eds) Rainfed agriculture: unlocking the potential. CABI International, Oxfordshire, U.K., pp 54–72

Pal DK, Bhattacharyya T, Chandran P, Ray SK, Satyavathi PLA, Durge SL, Raja P, Maurya UK (2009b) Vertisols (cracking clay soils) in a climosequence of Peninsular India: evidence for Holocene climate changes. Quatern Int 209:6–21

Pal DK, Bhattacharyya T, Wani SP (2012a) Formation and management of cracking clay soils (Vertisols) to enhance crop productivity: Indian Experience. In: Steward BA, Lal R (eds) World soil resources, Francis and Taylor, pp 317–343

Pal DK, Wani SP, Sahrawat KL (2012b) Vertisols of tropical Indian environments: pedology and edaphology. Geoderma 189–190:28–49

Pal DK, Bhattacharyya T, Sinha R, Srivastava P, Dasgupta AS, Chandran P, Ray SK, Nimje A (2012c) Clay minerals record from late quaternary drill cores of the Ganga Plains and their implications for provenance and climate change in the Himalayan Foreland. Palaeogeogr Palaeoclimatol Palaeoecol 356–357:27–37

Pal DK, Wani SP, Sahrawat KL, Srivastava P (2014) Red ferruginous soils of tropical Indian environments: A review of the pedogenic processes and its implications for edaphology. Catena 121:260–278. doi:10.1016/j.catena.05.023

Pal DK, Wani SP, Sahrawat KL (2015) Carbon sequestration in Indian soils: present status and the potential. Proc Natl Acad Sci Biol Sci (NASB), India, 85:337–358. doi:10.1007/s40011-014-0351-6

Rengasamy P, Sarma VAK, Murthy RS, Krishnamurti GSR (1978) Mineralogy, genesis and classification of ferruginous soils of the eastern Mysore plateau, India. J Soil Sci 29:431–445

Sanchez PA (1976) Properties and management of soils in the tropics. Wiley, New York

Sanchez PA, Logan TJ (1992) Myths and science about the chemistry and fertility of soils in the tropics. In: Lal R, Sanchez PA (eds) Myths and science of soils of the tropics. SSSA special publication number 29. SSSA, Inc and ACA, Inc, Madison, Wisconsin, USA, pp 35–46

Singer A (1980) The paleoclimatic interpretation of clay minerals in soils and weathering profiles. Earth-Sci Rev 15:303–326

Srivastava P, Pal DK, Aruche KM, Wani SP, Sahrawat KL (2015) Soils of the Indo-Gangetic Plains: a pedogenic response to landscape stability, climatic variability and anthropogenic activity during the Holocene. Earth Sci Rev 140:54–71. doi:10.1016/j.earscirev.2014.10.010

Tandon SK, Sinha R, Gibling MR, Dasgupta AS, Ghazanfari P (2008) Late quaternary evolution of the Ganga Plains: myths and misconceptions, recent developments and future directions. Golden Jubilee Memoir Geol Soc India 66:259–299

Thornbury WD (1969) Principles of geomorphology. Wiley Eastern Limited, New Delhi

Wilson MJ (1985) The mineralogy and weathering history of Scottish soils. In: Richards KS, Arnett RR, Ellis S (eds) Geomorphology and soils. George Allen and Unwin, London, pp 233–244

Chapter 2
Cracking Clay Soils (Vertisols): Pedology, Mineralogy and Taxonomy

Abstract A synthesis of recent developments in the pedology of Vertisols achieved through the use of high resolution micro-morphology, mineralogy, and age control data along with their geomorphic and climatic history, has contributed to our understanding of how the climate change-related pedogenic processes during the Holocene altered soil properties in the presence or absence of soil modifiers (Ca-zeolites and gypsum), calcium carbonate and palygorskite minerals. The climate change has caused modifications in the soil properties in the presence or absence of Ca-zeolites, gypsum, $CaCO_3$ and palygorskite minerals. The formation and persistence of Vertisols in the Deccan basalt areas under humid tropical (HT) climatic conditions, provides a unique example of tropical soil formation. Such soil formation remained incomprehensible unless the role of zeolites was highlighted by the Indian soil scientists during the last two decades. Persistence of these soils in HT climate for millions of years has provided a deductive check on the inductive reasoning of the conceptual models on the formation of Vertisols in HT climate. The novel insights will serve as guiding principles to improve and maintain their health and quality while developing suitable management practices to enhance and sustain their productivity. However, much of the success of the management interventions still depends on the proper classification of Vertisols at the subgroup level, identifying the impairment of drainage in Aridic Haplusterts (ESP ≥5, <15), Typic Haplusterts (with palygorskite) and the improvement of drainage in Sodic Haplusterts/Sodic Calciusterts with soil modifiers. The semi-arid tropics (SAT) Vertisols at present are less intensively cultivated because of their inherent limitations. It is hoped that new knowledge on pedology, mineralogy and taxonomy of dry and wet climates will fulfil the need for a handbook on Vertisols to facilitate their better management for optimizing their productivity in the 21st century.

Keywords Tropical soils · Vertisols · Pedology · Soil classification · Climate change

D.K. Pal, *A Treatise of Indian and Tropical Soils*,
DOI 10.1007/978-3-319-49439-5_2

2.1 Introduction

Vertisols are the most interesting and widely occurring soils of the world in general and the Indian subcontinent in particular. Because of their unique morphology these soils have attracted attention of both pedologists and edaphologists. This treatise is primarily based on the Indian Vertisols. However, relevant data from other tropical parts of the world are included. It uses state-of-the art data on the recent developments in the pedology of Vertisols, including variation in their morphological, physical, chemical, biological, mineralogical and micro-morphological properties, and aims to provide a better understanding of Vertisols created by the climate change phenomena of the Holocene. The updated information will facilitate in optimizing the efficient use and management of Vertisols in tropical India and other tropical regions.

Vertisols have attracted global attention in research, yielding a large body of data on their properties and management (Coulombe et al. 1996; Mermut et al. 1996). Despite the availability of substantial information on Vertisols; it still remains challenging to optimize their use and management (Coulombe et al. 1996; Myers and Pathak 2001; Syers et al. 2001; Pal et al. 2012a). The global area under Vertisols is estimated to be approximately 308 M ha, covering nearly 2.23% of the global ice-free land area (USDASCS 1994). But the reliability of this estimate still remains uncertain because several countries have not yet been included in the inventory (Coulombe et al. 1996). Moreover, the area under Vertisols in a soil survey may often be too small to resolve at the scale of map compilation. Such an example from the Indian sub-continent is shown in Table 2.1. Vertisols and vertic intergrades occur in 80 countries, but more than 75% of the global Vertisol area is contained in only 6 countries: India (25%), Australia (22%), Sudan (16%), the USA (6%), Chad (5%), and China (4%; Dudal and Eswaran 1988; Wilding and Coulombe 1996).

Vertisols occur in wide climatic zones, from the humid tropics to arid areas (Ahmad 1996), but they are most abundant in the tropics and sub arid regions. In the tropics, they occupy 60% of the total area; in the subtropics, they cover 30%, and they cover only 10% in cooler regions (Dudal and Eswaran 1988; Wilding and Coulombe 1996). In humid and sub-humid regions, Vertisols occupy 13% of the total land area; in sub-arid regions, 65%; in arid regions, 18%; and 4% in the Mediterranean climate, (Coulombe et al. 1996).

Vertisols are an important natural agricultural resource in many countries including Australia, India, China, the Caribbean Islands and the USA (Coulombe et al. 1996). Due to their shrink–swell properties and stickiness, Vertisols are known by a number of local regional and vernacular names (Dudal and Eswaran 1988). They are known in India by at least 13 different names (Murthy et al. 1982), which are related to the characteristic dark colour and/or to aspects of their workability. These soils are often difficult to cultivate, particularly for small farmers using handheld or animal-drawn implements since the roots of annual crops do not penetrate deeply due to poor subsoil porosity and aeration. This unfavourable physical condition of these soils compels farmers (especially in India) to allow these

Table 2.1 Distribution of Vertisols in different states of India under a broad bioclimatic systems

States	Bio-climate[a]	Area (m ha) (%)[b]
Uttar Pradesh	SAM, SHD	0.41 (0.12)
Punjab	SAM[d]	
Rajasthan	AD	0.98 (0.30)
Gujarat	AD, SAD, SAM	1.88 (0.57)
Madhya Pradesh	SAM,SHD, SHM[c]	10.75 (3.27)
Maharashtra	SAD, SAM, SHD, SHM[c]	5.60 (1.70)
Andhra Pradesh	SAD, SAM, SHD	2.24 (0.68)
Karnataka	AD, SHD, SHM, H	2.80 (0.85)
Tamil Nadu	SAD, SAM, SHD, SHM, H	0.91 (0.28)
Puducherry and Karaikal	SHM	0.011 (0.003)
Jharkhand	SHM, SHD	0.11 (0.034)
Orissa	SHM, SHD, H	0.90 (0.28)
West Bengal	SHD, SHM[d]	
Bihar	SHM[d]	
India		26.62 (8.10)

Adapted from Bhattacharyya et al. (2009)

[a]*AD* arid dry: 100–500 mm MAR (mean annual rainfall); *SAD* semi arid dry: 500–700 mm MAR; *SAM* semi arid moist: 700–1000 mm MAR; *SHD* subhumid dry: 1000–1200 mm MAR; *SHM* subhumid moist: 1200–1600 mm MAR; *H* Humid: 1600–2500 mm MAR

[b]Parentheses indicate percent of the total geographical area of the country

[c]In addition Vertisols occur in HT climate (>2500 mm MAR) in Madhya Pradesh and Maharashtra but they are not mappable in 1:250,000 scale (Bhattacharyya et al. 1993, 2005, 2009; Pal et al. 2012b)

[d]In the states of Punjab, Bihar, and West Bengal Vertisols and Vertic Intergrades also occur in SHM, SHD, and SAM climates (Pal et al. 2010) but they are not mappable in 1:250,000

soils to remain fallow during the rainy season and cultivate them only in the post-rainy season. Current agricultural land uses reflect a fact that although Vertisols are a relatively homogeneous soil group, they occur in a wide range of climatic environments globally and also show considerable variability in their uses and crop productivity (Pal et al. 2012b). As a matter of fact, Vertisol use is not confined to a single production system. In general, management of Vertisols is site-specific and requires an understanding of degradation and regeneration processes to optimize management strategies (Coulombe et al. 1996; Syers et al. 2001). It is often realized by the researchers (Puentes et al. 1988) that basic pedological research is needed to understand some of the unresolved edaphological aspects of Vertisols to develop optimal management practices.

Vertisols occur in wider bio-climatic zones in India (Table 2.1), in humid tropical (HT), sub-humid moist (SHM), sub-humid dry (SHD), semi-arid moist (SAM), semi-arid dry (SAD) and arid dry (AD) climatic environments. In total, they occupy 8.1% of the total geographical area of the Indian sub-continent (Table 2.1). Additionally, outside the Deccan basalt region of the Peninsula, in the states of Punjab, Bihar and West Bengal, Vertisols and their vertic inter grades occur in SHM, SHD and SAM climates (Pal et al. 2010), but they are not mappable at the 1:250,000 scale. By 2009, a total number of 306 BM (benchmark) Vertisols and vertic intergrades were identified by the ICAR-NBSS&LUP, Nagpur, India, which included 112 BM Vertisols (Pal et al. 2009a). This data set has been broadened to 425 BM soils by the National Agricultural Innovative Project (NAIP) (Component 4), sponsored research on 'Georeferenced soil information system for land use planning and monitoring soil and land quality for agriculture (GEOSIS) (www.geosis-naip-nbsslup.org).' through the Indian Council of Agricultural Research, New Delhi (Bhattacharyya et al. 2014; Mandal et al. 2014). Research scientists of Division of Soil Resource Studies of the ICAR-NBSS&LUP and GEOSIS team members during the last two and half a decade examined more than 200 Vertisols in the states of Madhya Pradesh, Maharashtra, Chhattisgarh, Karnataka, Andhra Pradesh, Tamil Nadu, Gujarat, Rajasthan and West Bengal in India. They have been indicated (along with their global distribution) on a 1:1 million-scale map (NBSS&LUP 2002; Pal et al. 2012a). Further, based on a larger data set (Mandal et al. 2014) a fresh initiative by the GEOSIS developed a revised map (on a 1:1 million-scale) of the black soil region (BSR) of the Indian sub-continent, which shows the presence of Vertisols.

2.2 Factors of Vertisol Formation: A Revisit

The soil-forming factors are the most relevant and appropriate features that explain Vertisol formation. They are interdependent and highly variable and therefore influence the properties of Vertisols in multiple ways as described and explained in several text books on soil science. However, in view of the recent developments in studies of Vertisol formation in the Indian sub-continent Pal et al. (2001, 2006a, b, 2009a, b, c, 2012a, b, c; Bhattacharyya et al. 2005; Ray et al. 2006a; Srivastava et al. 2002, 2010, 2015), each of these five factors merits revisit and discussion.

2.2.1 Parent Material

The basic parent materials, essentially required for the formation of Vertisols, are made available through several geologic formations (Murthy et al. 1982). The parent materials from inheritance or weathering provide a large quantity of smectites but the distinction between inherited and newly-formed clay minerals is difficult to discern. However, in Vertisols of the sub-humid, semi-arid and arid climates of Peninsular India, chemical weathering of primary minerals is not substantial as evidenced from the presence of either fresh or weakly to moderately altered plagioclase and micas (Fig. 2.1a, b). Such stages of mineral weathering discount the formation of smectite during the development of Vertisols (Srivastava et al. 2002) and thus validate the hypothesis that Vertisol formation reflects a

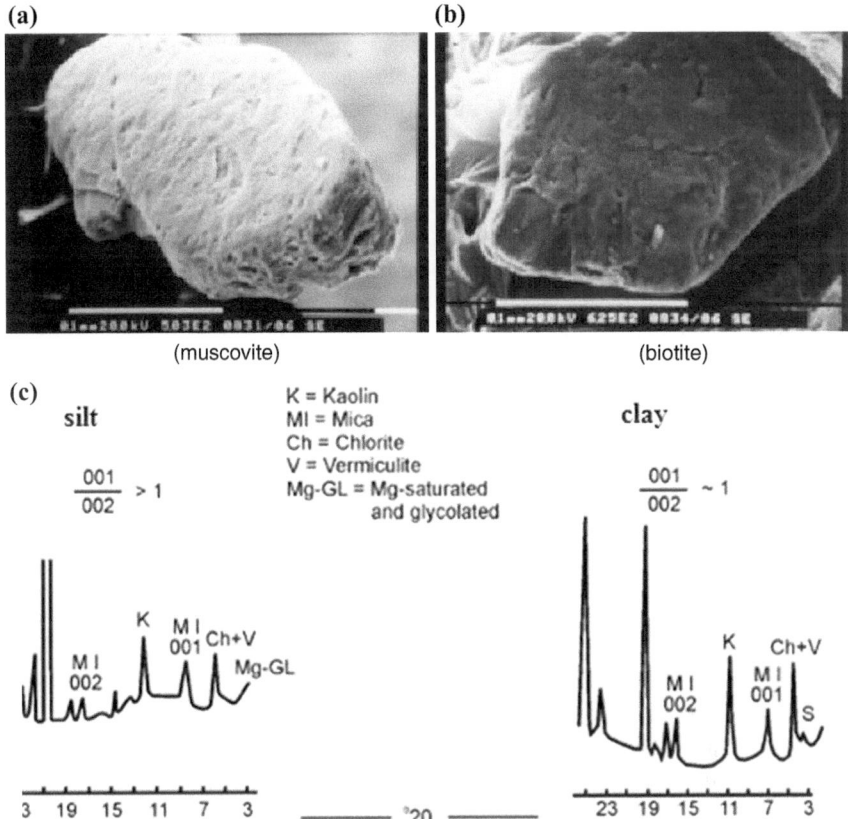

Fig. 2.1 Representative SEM photographs showing no or very little alteration of micas (**a, b**), and XRD diagrams of the silt and clay fractions (**c**) of Vertisols of Peninsular India (Adapted from Pal et al. 2006c)

positive entropy change (Smeck et al. 1983). Smectite clay is solely responsible for the shrinking and swelling phenomena to create vertic properties of soils (Shirsath et al. 2000). Therefore, smectite is the exclusive parent material for the formation of Vertisols.

2.2.2   Climate

Vertisols occur from the humid tropics to arid areas (Ahmad 1996). Although their characteristics are often related to overall climate, other factors such as texture, clay mineralogy, cation saturation, and the amount of exchangeable sodium equally influence soil morphology (Dudal and Eswaran 1988; Eswaran et al. 1988; Pal et al. 2012a). But the formation of Vertisols in humid and arid bio-climatic environments is difficult to follow because large quantities of smectite are required to create their shrink–swell properties. Smectite is ephemeral in an HT climate (Bhattacharyya et al. 1993; Pal et al. 1989) while in sub-humid to arid climates, the weathering of primary minerals contributes very little to smectite formation (Srivastava et al. 2002). Thus smectite clay minerals cannot be retained in HT Vertisols while they are in huge quantity in Vertisols of sub-humid to arid climatic conditions. The occurrence of Vertisols in the alluvium of weathering Deccan basalt in HT, SHM, SHD, SAM, SAD, and AD environments of the Indian peninsula (Pal et al. 2009a), may suggest that the basaltic parent material influenced soil formation such that Vertisols are formed under different climatic conditions (Mohr et al. 1972). It is interesting to note that the morphological and chemical properties of these Vertisols differ. In general, the colour of soils of the HT climate is dark brown (7.5YR 3/3) to dark reddish (5YR 3/3) and yellowish brown (10YR 3/4) and it is dark (10YR 3/1) to very dark grayish brown (10YR 3/2) in soils of other climates. The subsoils of the HT climate have weak and small wedge-shaped aggregates with pressure faces that break to weak angular blocky structure whereas those of SHM, SHD, SAM, SAD and AD climates have strong medium sub-angular blocky to strong coarse angular blocky structure with pressure faces and slickensides that break into small angular peds. More interestingly, cracks >0.5 cm wide extend down to the zones of sphenoids and wedge-shaped peds with smooth or slickensided surfaces in HT, SHM, SHD and SAM soils, but cracks cut through these zones in SAD and AD soils (Fig. 2.2). Soil reactions and the $CaCO_3$ content indicate that a reduction in mean annual rainfall (MAR) leads to the formation of calcareous and alkaline soils. Hence, the soils are Typic Haplusterts in HT, Typic/Udic Haplusterts in SHM, SHD and SAM climates, and Sodic Haplusterts and Sodic Calciusterts in SAD and AD climates (Pal et al. 2009a). In view of these observations, the hypothesis of Mohr et al. (1972) for the formation of similar soils in basaltic alluvium under different climates is inadequate to explain the formation of Vertisols of tropical India. The abundance of Vertisols in dry climates may suggest a role of climate in their genesis (Eswaran et al. 1988) but it will be more appropriate to follow their genesis as influenced by a specific bio-climatic environment. Occurrence of Vertisols in a climosequence provides evidence for the Holocene climate changes in tropical and subtropical regions of India and elsewhere (Pal et al. 2009a).

2.2.3 Topography

Despite the fact Vertisols occur abundantly in low elevations they do occur at higher elevations in the Ethiopian plateau or on higher slopes, as in the West Indies (Coulombe et al. 1996) and in micro-depressions on the Deccan basalt plateau in the Indian sub-continent (Bhattacharyya et al. 1993; Pillai et al. 1996). The majority of the Indian Vertisols occur in lower physiographic areas, i.e., in the lower piedmont plains and valleys (Pal and Deshpande 1987; Pal et al. 2009a). Vertisols in micro-depressions on the Deccan basalt plateau are spatially associated with red

Panjri, Nagpur, Maharashtra.	Nipani, Adilabad, Andhra Pradesh	Bhatumbra, Bidar, Karnataka	Paral, Akola, Maharashtra	Sokhda, Rajkot, Gujarat
Typic Haplusterts SHM, MAR 1127mm	Typic Haplusterts SHD, MAR 1071 mm	Udic Haplusterts SAM. MAR 977 mm	Sodic Haplusterts SAD. MAR 794 mm	Sodic Calciusterts AD. MAR 533 mm

Fig. 2.2 Cracks are extending beyond the zone of slickensides with increase in aridity (SHM to AD bioclimates) (Adapted from Pal et al. 2003a)

ferruginous soils (Alfisols) and exist as distinct entities under almost similar topographical conditions in the HT (Bhattacharyya et al. 1993) and SAD (Pillai et al. 1996) climates. The red soils are mildly acidic Entisols/Inceptisols/Alfisols in SAD climates, but they are moderately acidic Alfisols in HT climate. The red soil clays of an HT climate contain predominant amount of smectite–kaolinite (Sm–K), whereas those of the SAD climate contain only small amounts of Sm–K. The clay Sm–K was formed at the expense of smectite in red HT soils, and in red SAD soils, it is considered to have originated under a previous, humid climate regime (Pal 2003). Formation of both red soils (Alfisols) and Vertisols in the contrasting climate has been explained through the landscape reduction process (Bhattacharyya et al. 1993; Pal 1988), as in similar soils elsewhere (Beckman et al. 1974). It is believed that in the initial stage of soil formation, weathering products rich in smectite from the hills were deposited in micro-depressions, as is evident from the lithic/paralithic contacts of such Vertisols (Fig. 2.3a). Over time, the hill sites gradually flattened, and internal drainage dominated over surface run-off. After peneplanation, the red soils (Alfisols) of the present (Bhattacharyya et al. 1993) and the past (Pillai et al. 1996) HT climates on relatively stable surfaces continued to weather to form the Sm–K. The spatially-associated Vertisols, however, continued to exist in the micro-depressions (Fig. 2.4) even in HT climates because the smectite was

(a)   **(b)**

Fig. 2.3 A representative Vertisol (Typic Haplusterts) developed in micro-depression of a plateau, showing paralithic contact with the Deccan basalt of central India (**a**) and in the Purna Valley developed in the alluvium of the Deccan basalt (**b**). *Photograph* courtesy—DKP

stabilized due to the continuous supply of bases from Ca-rich zeolites (Bhattacharyya et al. 1993). During the Plio-Pleistocene transition, the period of the HT climate ended (Pal et al. 1989), and both smectite and Sm–K in SAD Vertisols were preserved to the present. The reduced rainfall in SAD climate restricted further leaching in Vertisols and caused calcareousness and the rise in pH (Pillai et al. 1996). Thus, Vertisols are not common in residuum on plateau of the Deccan basalt, and also in the valley of the Deccan basalt areas (Fig. 2.3b).

Another form of micro-topography "gilgai" in Vertisol areas is not very well understood and at present, gilgai micro-topographies are very rare on the Indian sub-continent because most were obliterated by post-cultural human activities. Wilding and Coulombe (1996) reported that the depth distribution of soil properties generally differs between the mounds and depressions of the gilgai topography, which results in vertical and horizontal spatial variability in Vertisols within distances as short as a few metres or less. Such spatial and horizontal variability in SAD Vertisols is common in a central Indian watershed (Vaidya and Pal 2002). Vertisols in the watershed occur in both micro-high (MH) and micro-low (ML) positions. The distance between these positions is approximately 6 km, and the elevation difference is 0.5–5 m (Fig. 2.5). MH Vertisols are more clayey and strongly calcareous, alkaline and have poor drainage (saturated hydraulic conductivity, sHC < 10 mm/h), and those in ML positions are less clayey and calcareous, mildly alkaline and have better drainage (sHC > 10 mm/h) (Table 2.2). MH

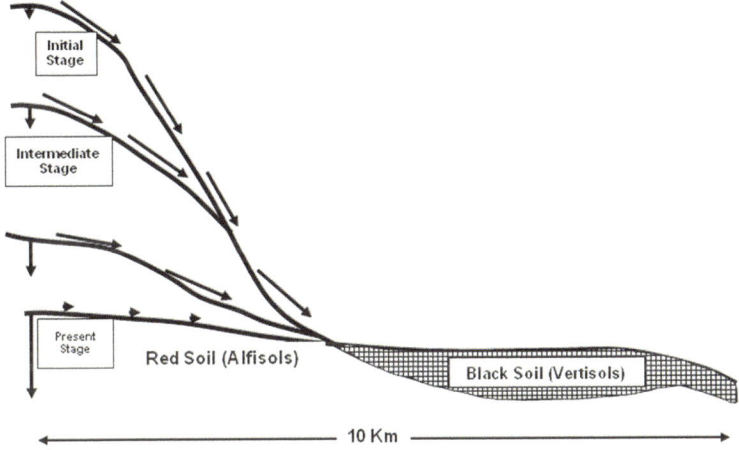

Fig. 2.4 Schematic diagram of the pedon site of red soils (Alfisols) and black soils, (Vertisols) showing the landscape reduction process explaining the formation of spatially associated red and black soils (Adapted from Pal 2008)

Fig. 2.5 Juxtaposition of the occurrence of sodic and non-sodic soils (Vertisols) on MH and ML positions in black soils region (Adapted from Vaidya and Pal 2002)

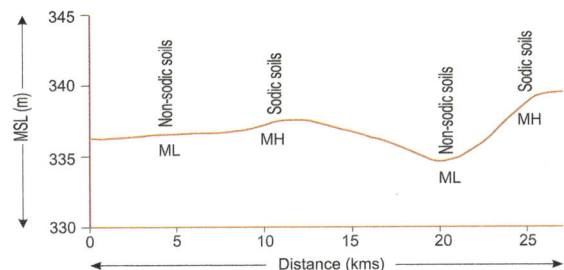

Vertisols have larger amounts of $CaCO_3$ because of their relatively more aridic environment than in ML Vertisols. During the winter season, moisture of sodic soils of the Bss horizons of the MH positions is held at 300 kPa while it is between 33 and 100 kPa in non-sodic soils of the ML positions (Kadu 1997; Deshmukh et al. 2014), confirming that the soils of the MH positions remain relatively more arid. During the very hot summer months, this results in much less water in the subsoils of the MH positions where deep cracks cut through the Bss horizons. The lack of adequate soil water during the shrink–swell cycles restricts the swelling of smectite and results in weaker plasma separation in soils of the MH positions. The ML Vertisols showed strong plasma separation with parallel/porostriated b-fabric (Fig. 2.6a) whereas the MH Vertisols showed weak plasma separation with stipple-speckled to mosaic-speckled plasmic fabric (Fig. 2.6b). The plasma separation is more pronounced and the preferred orientation in zones adjacent to grains and voids is stronger in soils of ML than those of the MH positions. Thus, although surface-oriented plasma separation of soils indicates a high degree of clay activity

Table 2.2 Physical and chemical properties of Vertisols on MH and ML positions in Pedhi Watershed of Maharashtra

Horizon	Depth (cm)	pH (1:2) H_2O	ECe (dS m^{-1})	Organic carbon (%)	$CaCO_3$ < 2 mm (%)	Clay %	sHC $(mm/h)^a$	ESP^b
Representative sodic Vertisols of MH position: Pedon 6: Khartalegaon: Sodic Haplusterts							
Ap	0–15	8.1	0.4	0.9	13.3	54	7.7	2.0
Bw1	15–37	8.3	0.5	0.7	11.9	58	6.5	3.5
Bw2	37–55	8.5	0.9	0.7	11.8	66	1.0	8.8
Bss1	55–86	8.6	0.7	0.5	11.2	64	0.9	10.2
Bss2	86–136	8.8	1.0	0.4	11.4	64	0.8	16.8
Bss3	136–150	8.8	2.4	0.4	15.2	64	0.5	20.2
Representative non-sodic Vertisols of ML position: Pedon 2: Wadura: Aridic Haplusterts							
Ap	0–13	8.1	0.8	0.9	5.8	82	7.2	4.1
Bw1	13–36	8.1	0.8	0.6	6.4	76	12.4	4.3
Bw2	36–58	8.1	0.7	0.8	6.6	79	16.4	4.1
Bss1	58–87	8.0	0.5	0.8	6.7	81	29.8	2.6
Bss2	87–125	8.0	0.5	0.8	6.4	86	21.2	1.8

Adapted from Vaidya and Pal (2002), ^{a}Saturated hydraulic conductivity, ^{b}Exchangeable sodium percentage

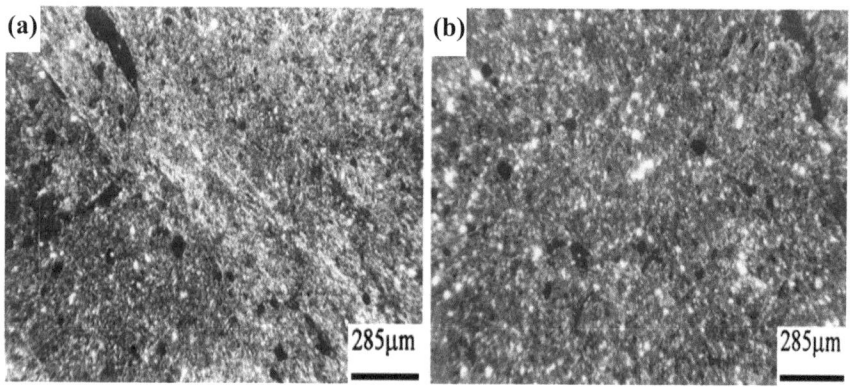

Fig. 2.6 Representative photograph of plasmic fabric in cross polarized light. **a** Strong plasma separation with parallel-striated b fabric, **b** weak plasma separation with mosaic–speckled plasma separation (Adapted from Vaidya and Pal 2002)

and shrink–swell, the plasmic fabric is not uniform among the soils of the ML and MH positions. The formation of sodic soils in MH positions alongside nonsodic soils in ML positions is a unique phenomenon and is described in further details under soil degradation in Chap. 6.

2.2.4   Vegetation

Influence of vegetation on pedogenesis and distribution of Vertisols in the world are rarely observed (Coulombe et al. 1996). Vertisols that are not cultivated are associated with native vegetation, such as grasslands and savannahs (Probert et al. 1987). Vertisols can tilt large trees (Bhattacharyya et al. 1999). Not surprisingly, few, if any, commercial forests are found on Vertisols (Buol et al. 1978), but mixed pine and deciduous forests are reported in selected regions of east Texas. Most Vertisols at present are under post-cultural activities and that make it difficult to identify and infer the influence of native vegetation (Coulombe et al. 1996).

In India, Vertisols are often difficult to cultivate because of their poor subsoil porosity and aeration. As a result roots of annual crops do not penetrate deeply and farmers allow these soils to lie fallow for one or more rainy seasons or cultivate them only in the post-rainy season. Thus, Vertisols have limitations that restrict their full potential to grow during both rainy season and winter crops and generally are less intensively cultivated (Pal et al. 2012a). Thus the present management interventions have a minimal role in their formation and modification. Vertisols have low organic carbon status on both the surface and sub-surface layers (<1%) and a moderate to high content of $CaCO_3$, indicating that biotic factors have no substantive role in the genesis of Vertisols (Pal et al. 2009a, b).

2.2.5   Time

Almost all Vertisols are derived from geological rock systems that are millions of years old but their real age is not so old (Pal et al. 2012a). The age of the parent material provides only a maximum chronological point. In reality, the true age of the geomorphic surface or the time required for Vertisol formation is much less than millions of years (Pal et al. 2012a). Many pedologists suggest that the formation of slickensides (the most essential physical characteristic to qualify for vertic properties) is very rapid and that Vertisols are formed on geomorphic surfaces in as few as 550 years (Parsons et al. 1973). For example, a Vertic Haplustalf formed <100 year. BP (^{14}C age, Pal et al. 2006a) in the Deccan basalt alluvium of the central Indian SHD bio-climate, exhibits pressure faces but lacks slickensides. The occurrence of shrink–swell soils and slickensides in the SHM lower Indo Gangetic Plains of 500–1500 year. BP (TL ages) indicates that a minimum of 500 year. is adequate to form Vertisols (Singh et al. 1998). The Vertisols of central and western peninsular India developed in the Deccan basalt alluvium of the Upper Cretaceous, mostly during the Holocene period are, however, older than this age; they have a minimum ^{14}C age of 3390 year, and a maximum of 10,187 year. BP (Pal et al. 2006a, 2009a). These data suggest that Vertisols in India and elsewhere are developed during the Holocene period.

2.3 Smectite Clay Minerals in Vertisols: Recent Initiatives for Their Proper Characterization

To establish a link between smectitic mineralogy and important bulk properties of Vertisols especially their shrink-swell characteristic, often becomes difficult because on many occasions the description of smectitic minerals is inadequate or incomplete. As soil smectites differ from "type" minerals, it becomes essential to identify and characterize them properly as they occur in soil systems. Although the Vertisol clays are in general a mixture of several other clay minerals, adequate description of clay smectite is possible with sustained efforts and such exercises made in recent years, are illustrated in the following.

2.3.1 Characteristics of Smectite

The shrinking and swelling of soils is primarily governed by the nature of the clay minerals, particularly their surface properties. The soils containing all other clays shrink and swell with variations in moisture content but changes are particularly extreme in smectites (Borchardt 1989). While relating the shrink-swell properties and clay mineral types Bhattacharyya et al. (1997) concluded that the vertic properties of soils are a function of smectite content. Despite this basic understanding, several non-expanding clay minerals (kaolin, micas, chlorites, palygorskite and vermiculites) are often mentioned to be associated with shrink-swell properties of Vertisols and their vertic intergrades (Hajek 1985; Coulombe et al. 1996; Heidari et al. 2008). In addition, kaolinite is reported to be abundant in some Vertisols in El Salvador (Yerima et al. 1985, 1987) and Sudan (Yousif et al. 1988) and given secondary importance in Vertisols of the USA (Hajek 1985). Many such soils of the USA do have clay CEC > 40 and those of El Salvador show higher clay CEC (62–79 cmol (+)/kg) and COLE (0.10–0.12). Such high clay CEC values do indicate that the presence of expansible minerals might have escaped the notice of researchers in the few shrink–swell soils of the USA (Hajek 1985) and El Salvador (Yerima et al. 1985, 1987). However, a close examination of the X-ray diffraction (XRD) diagrams of the fine clays of El Salvador soils in which shrink–swell processes are related to the fine clay kaolin content (Yerima et al. 1985, 1987) indicates the presence of a smectite peak in the Atiocoyo soils. Moreover, soils also have dominating amount of Sm–K, which is capable of inducing the vertic character in soils (Bhattacharyya et al. 1993).

Soil Survey Staff (1994) stipulated the montmorillonitic mineralogy of soils is associated with vertic properties when smectite exceeds 50% of the total mineral content in the <2 μm clay fraction. Later on, a qualitative smectite mineralogy class was proposed by the Soil Survey Staff (1998, 1999) for the soils that contain more smectite by weight than any other single clay mineral. This requirement provided a means by which smectite can reflect a quantitative dimension of the vertic properties of soils. Quantitative determination of minerals in the soil clay fractions by XRD analysis is difficult, as any attempts in this regard have yielded semi-quantitative estimates. Moreover, such estimation is not infallible when minerals are in the interstratified phase. The presence of Sm–K in shrink–swell soils is common in India and elsewhere (Pal et al. 2012a). Although the peak-shift analysis (Wilson 1987) is a useful method to determine the smectite content in Sm–K, it becomes ineffective when the smectite component in Sm–K is highly chloritised and the swelling of smectites on glycolation is restricted. To circumvent this problem, the chemical method of Alexiades and Jackson (1965) is an effective way to quantitatively determine the smectite content in soil clays and thus, Shirsath et al. (2000) observed a strong relationship between marked shrink–swell properties and smectite content in the clay fraction (<2 μm). Vertic properties with a linear extensibility (LE) of 6 in shrink–swell soils correspond to a minimum threshold value of 20% clay smectite, therefore suggesting that only smectitic soils should be considered shrink–swell soils in the US Taxonomy. Of the three smectite species (montmorillonite, beidellite and nontronite), montmorillonite and beidellite are the most commonly reported in Vertisols, while reports of nontronite are rare (Coulombe et al. 1996). In the Indian sub-continent, the majority of shrink–swell soils are developed in the alluvium of weathering Deccan Basalt, covering an area of 500,000 km^2 (Duncan and Pyle 1988). A review on the mineralogy of Indian shrink–swell soils (Vertisols and vertic intergrades) (Ghosh and Kapoor 1982) indicates that the soil clays are dominated by beidellite–nontronite type minerals. These authors reported the results of computed clay minerals based on smectite formulae, although such an approach is not infallible (Sawhney and Jackson 1958). X-ray diffraction analysis of large numbers of smectite dominated fine clays of Indian shrink–swell soils (Pal 2003) indicates the presence of small to moderate amounts of hydroxy-interlayer (HI) material in the smectite interlayers, alongside a small amount of vermiculite. Both hydroxy-interlayers and vermiculite are not easily detected in the glycolation samples but are discernible during gradual heating from 110 to 550 °C of the K-saturated samples. Hydroxy-interlayered smectite is detected from low-angle-side broadening of the 1.0-nm peak at 550 °C (Wildman et al. 1968; Fig. 2.7) while vermiculite is detected when the 1.0-nm peak of mica is reinforced on heating to 110 °C (Pal and Durge 1987). Even a small quantity of such impurities (HI materials and vermiculite) affects the charge and sum relationships using smectite formulae.

The presence of both montmorillonite and beidellite in the fine-clay fractions of Indian shrink–swell soils in basaltic alluvium was confirmed by the Greene-Kelly test (Greene-Kelly 1953), and the former dominates over the latter (Pal et al. 2012a). It is interesting to note that on glycerol vapour treatment (Harward et al. 1969), the clay smectites expand to approximately 1.9 nm, indicating only the presence of montmorillonite (Fig. 2.7). In other words, fine-clay smectite is nearer to montmorillonite in the montmorillonite–beidellite series. Since the nontronite would behave like beidellite in these tests and clay smectite is unstable under HCl treatment, consequently releasing considerable iron in solution, it was concluded that the smectite in Vertisols is nearer to the montmorillonite of the

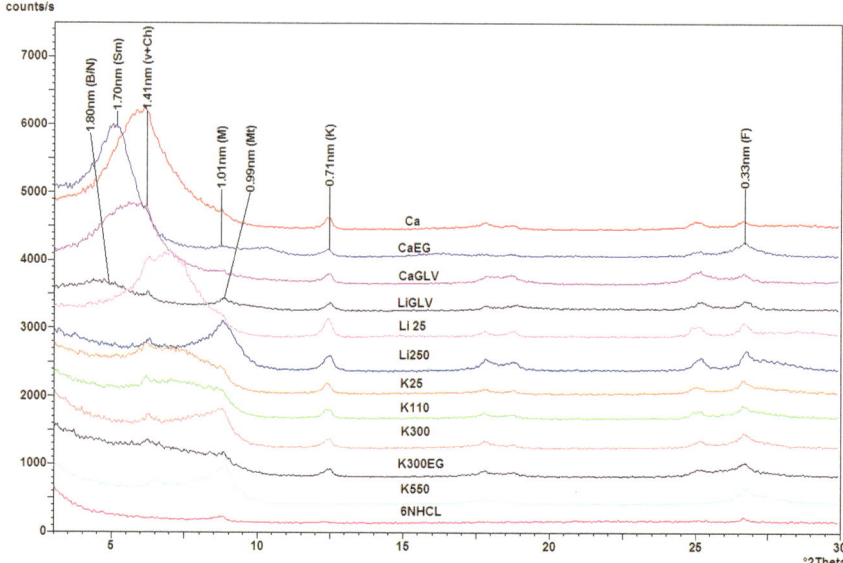

Fig. 2.7 Representative X-ray diffractograms of fine clay fractions (<0.2 μm) of the Bss horizons of Vertisols of central India; Ca = Ca saturated; Ca-EG = Ca saturated plus ethylene glycol vapour treated; CaGLV = Ca-saturated plus glycerol vapour treated; Li = Li-saturated and heated to 25, 250 °C (16 h), LiGLV 30-D = Li-saturated and heated at 250 °C plus glycerol vapour treated and scanned after 30 days; K25/110/300/550 °C = K-saturated and heated to 25, 110, 300, 550 °C; K300EG = K-saturated and heated to 300 °C plus ethylene glycol vapour treated; 6NHCl = 6N HCl treated fine clays; *Sm* Smectite, *B/N* beidellite/nontronite; *V + Ch* vermiculite plus chlorite; *M* mica; *Mt* montmorillonite; *K* Kaolinite; *F* Feldspars (Adapted from Pal et al. 2003b; Bhople 2010)

montmorillonite–nontronite series (Pal and Deshpande 1987). Smectite expands beyond 1.4 nm after glycolation of the K-saturated and heated samples (300 °C; Fig. 2.7). This expansion behaviour indicates its low-layer charge density as of specimen dioctahedral smectite, which is also evidenced by its no K-selectivity (Pal and Durge 1987).

2.3.2 Seat of Charge in and Layer Charge of Fine Clay Smectites

Although some knowledge on low charge density of clay smectites is gained as discussed above, determination of their layer charge always remains a fundamental requirement to all physical and chemical properties of soils such as soil structure, drainage, aeration, water retention (Laird et al. 1988), cation exchange reactions, specific surface area and degree of hydration (Wilding and Tessier 1988). Ideally, the charge in the layer silicate minerals should be either in the tetrahedral sheet or octahedral sheet, but it is usually observed that the charge is distributed over both the sheets (Malla and Douglas 1987). It is essential to locate seat of charge and also to study changes in the proportion of tetrahedral and octahedral charge during the pedogenetic processes of soil formation. Information about the position of charge in smectites of Indian Vertisols was made available by some researchers (Kapse et al. 2010; Bhople et al. 2011) who made attempts to locate the seat of charge of some selected bench mark Vertisols' fine clay smectites of central India by using cation exchange capacity (CEC) method of fine clays with the aid of mechanism of charge reduction advocated by Hofmann and Klemen (1950). Their results indicate that CECs are distributed in both tetrahedral and octahedral layers of which the contribution of the former is higher (>50%) than the latter (Table 2.3). These researchers suggested that the determination of reduced CECs from Greene-Kelley test (Greene-Kelly 1953) is an effective means of measuring the octahedral and tetrahedral CECs and also for calculating the charge of soil clays.

The layer charge also determines the properties of soil clay minerals and indicates a mineral's capacity to retain cations and adsorb water and other polar organic molecules (Malla and Douglas 1987).It is also known that different swelling properties of clays with identical interlayer cations are mainly due to differences in layer charge densities (Weiss et al. 1955). Thus, it is important to determine the layer charge of smectite minerals. Theoretically, this parameter should range between 0.3 and 0.6 electrons per half unit cell in smectites. Tessier and Pedro (1987), however, reported that high-charge smectite (between 0.45 and 0.60

Table 2.3 CECs (total, tetrahedral and octahedral) and the contribution of tetrahedral and octahedral CECs to total CEC of fine clay smectites of some representative benchmark Vertisols of central India

Horizon	Depth (cm)	CEC total cmol $(p+)kg^{-1}$	CEC$_{Tetrahedral}$ cmol $(p+)$ kg^{-1}	CEC$_{Octahedral}$ cmol $(p+)$ kg^{-1}	Contribution of CEC$_{Tetrahedral}$ to CEC$_{Total}$ (%)	Contribution of CEC$_{Octahedral}$ to CEC$_{Total}$ (%)
Linga Series: Nagpur: Maharashtra: Typic Haplusterts						
Ap	0–16	83	68	15	82	18
Bw1	16–44	75	59	16	78	22
Bw2	44–69	80	52	28	65	35
Bss1	69–102	84	54	30	64	36
Bss2	102–128	93	48	45	52	48
Bss3	128–150	84	49	35	58	42
Nimone Series: Ahmadnager: Maharashtra: Sodic Haplusterts						
Ap	0–13	59	58	1	98	2
Bw1	13–38	66	46	20	69	31
Bw2	38–55	72	47	25	65	35
Bss1	55–94	66	41	25	62	38
Bss2	94–128	54	38	26	70	30
Bw/Bc	128–150	59	39	20	66	34

Adapted from Kapse et al. (2010), Bhople et al. (2011)

electrons per half unit cell), is common in soils. Several researchers (Bardaoui and Bloom 1990; Chen et al. 1989) also reported the presence of smectite in Vertisols with a layer charge in the range of vermiculite (0.6–0.9 electrons per half unit cell). The smectite charge in selected bench-mark Indian Vertisols is distributed in both tetrahedral and octahedral layers and also showed a high layer charge (0.28–0.78 mol electrons/$(SiAl)_4O_{10}(OH)_2$), and low-charge smectite constitutes >70% in them (Ray et al. 2003); however, a charge >0.6 was attributed to the presence of a small quantity of vermiculite (5–9%, Pal and Durge 1987) and to the presence of hydroxy-interlayering in smectite interlayers (i.e. hydroxy-interlayered smectites, HIS) (Ray et al. 2003). Conversely, in some Vertisols of central India developed in Deccan basalt alluvium where hydroxy-interlayering in the smectite interlayers is negligible, especially in Sodic Haplusterts, the layer charge of fine clay smectites showed a lowest value of 0.307–0.353 and 0.328–0.360 mol(−)/$\{(Si, Al)_4O_{10}(OH)_2\}$ for Paral (Sodic Haplusterts) and Boripani (Typic Haplusterts) soils in the state of Maharashtra of central India, respectively. These values are nearer to the layer charge of bentonite (Wyoming) having 0.26 mol(−)/$\{(Si, Al)_4O_{10}(OH)_2\}$ (Thakare et al. 2013). The layer charge of clay smectites in Indian Vertisols by the alkyl ammonium method (Lagaly 1994) showed the presence of monolayer to bilayer and bilayer to pseudotrilayer transitions, indicating heterogeneity in the layer-charge density (Fig. 2.8) (Ray et al. 2003; Bhople 2010; Thakare et al. 2013).

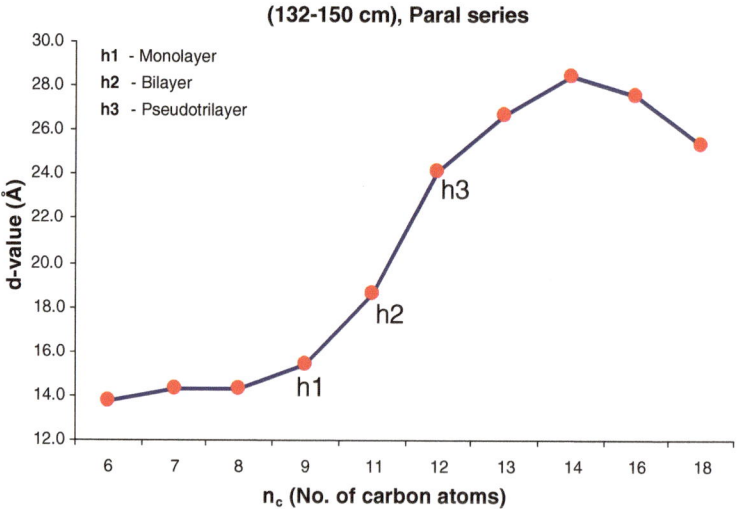

Fig. 2.8 Representative S-type curve relationship between d-spacings (001) of fine clay smectites intercalated with alkylammonium chlorides and number of carbon atoms, indicating the heterogeneity in layer charge of fine clay smectites of Indian Vertisols (Adapted from Bhople 2010)

2.3.3   Hydroxy-Interlayered Smectites (HIS) and Determination of Layer Charge

Although the occurrence of HIS is very common in Vertisols of peninsular India, the extent of hydroxy-interlayering however varies. In some soils it is in negligible amounts (Thakare et al. 2013) but the presence of low to moderate hydroxy-interlayering is more common (Pal 2003; Pal and Deshpande 1987). Ray et al. (2002) observed that higher the tetrahedral charge; greater is the probability of hydroxy-interlayering in fine clay smectites. But the Vertisols' fine clay smectites do have higher tetrahedral charge than in octahedral layer (Kapse et al. 2010; Bhople et al. 2011). Therefore, the precise reason for such variation in amount of hydroxy-interlayering is yet to be resolved. Hydroxy-Al interlayer clays give 2:1 layer phyllosilicate great weathering resistance, and the interlayered clays are an important component of both moderately weathered and intensively weathered soils. The hydroxy interlayers prevent the determination of the layer charge by the alkylammonium method (Lagaly 1994) by obstructing the normal intrusion of alkylammonium ions into interlayers (Ray et al. 2006b). Such interlayers result in the formation of relatively more paraffin type layers with bi-pseudotrilayer transition (Fig. 2.8), which causes the overestimation of the layer charge of core lattice mineral. In order to determine the layer charge of the core lattice minerals, Ray et al. (2006b) used 0.25 N EDTA solutions (pH 7.0) to remove the HI materials from the fine-clay smectites of Indian Vertisols to determine the layer charge of the cleaned clays and observed the removal of HI materials by the EDTA solutions was almost complete. Ray et al. (2006b) obtained the weighted-average layer charge of the pre-treated clays from 0.40 to 0.46 $mol(-)/(Si, Al)_4O_{10}(OH)_2$, and after EDTA treatment, the charge ranged from 0.27 to 0.33 $mol(-)/(Si, Al)_4O_{10}(OH)_2$, a value range close to the layer charge of Wyoming bentonite (montmorillonite species of the smectite group of layer silicate minerals) (Thakare et al. 2013). These authors also observed that after EDTA treatment, the Ca-saturated and glycolated fine clays showed greater X-ray intensity than their corresponding Ca-treated curves. The K-treated curves also showed marked differences when compared with the original untreated fine clays. The improvement in the intensities of glycolated and K-saturated samples showed the effectiveness of EDTA solutions in the removal of HI materials and also in determining the actual layer charge in soil-clay smectites.

2.4   Genesis of Smectites in Vertisols: A Revised Status

It is observed that smectite is the most abundant and essential phyllosilicate in Vertisols worldwide, and it remains either as a discrete mineral or as a mineral interstratified with any other layer silicates. A minimum of 20% smectite in their clay fractions (<2 μm; Shirsath et al. 2000) is required for the manifestation of vertic properties at a linear extensibility (LE) of 6 in shrink–swell soils.

Smectite-clay minerals are ephemeral in the HT climate, where they are rapidly transformed to kaolin (a 0.7 nm mineral consisted of hydroxy-interlayered smectite and kaolinite, Sm-K). Therefore, it is difficult to understand the formation of Vertisols in HT climates. Recent research has explained their formation with the presence of soil modifiers such as Ca-zeolites (Bhattacharyya et al. 1993; Pal et al. 2006a) that release Ca^{2+} ions to prevent the complete transformation of smectite to kaolin, and the high base status helps in stabilizing and retention of smectite. Therefore, the formation and persistence of slightly acidic to acidic Typic Haplusterts with predominant Sm–K in clay fractions in India (Bhattacharyya et al. 1999, 2005; Pal et al. 2009a) and elsewhere (Ahmad 1983) is possible only in presence of soil modifiers that maintain the base saturation well above 50% (Pal et al. 2006b, 2013a).

Despite the fact that Vertisols are abundant in semi-arid regions (Eswaran et al. 1988), the large quantities of dioctahedral smectite required for the formation of Vertisols cannot form in semi-arid soils, as the primary minerals contribute little towards the formation of smectites in the prevailing dry climates (Srivastava et al. 2002; Pal et al. 2009a). XRD analysis of fine clays in Indian SHM, SHD, SAM, SAD and AD Vertisols indicates that dioctahedral smectites are fairly well crystallized, as they yield sharp basal reflections on glycolation and show regular higher (though short and broad) reflections, and show no sign of transformation except for the low to moderate amounts of hydroxy interlayering in the smectite interlayers (Pal et al. 2009a). Hydroxy-interlayering is also observed in the silt and coarse clay sized vermiculite (HIV) that resulted in the formation of pseudo- or pedogenic chlorite (PCh; Pal et al. 2012b; Fig. 2.9). The presence of fine clay size hydroxy-interlayered dioctahedral smectite (HIS), as well as the silt and coarse clay sized HIV and PCh indicates that the hydroxy-interlayering in the vermiculite and smectite did not occur in the present slight to moderate alkaline soil reaction induced by dry climates. It can only occur when positively charged hydroxy interlayer materials (Barnhisel and Bertsch 1989) entered into the inter-layer spaces in acidic soil pH (<6.0) (Rich 1968). The majority of Vertisols in sub humid to arid climates all over the world have pH values either near to neutral or well above 8.0 throughout. Under such alkaline soil reaction, the 2:1 layer silicates suffer congruent dissolution (Pal 1985), and thus discounts the hydroxy interlayering of smectites after deposition of the basaltic alluvium (Pal et al. 2012b). The hydroxy-interlayering in vermiculite and smectite and the subsequent transformation of vermiculite to PCh do not, therefore, represent contemporary pedogenesis of Vertisols in dry climates. Indian Vertisols contain both NPC (relict Fe–Mn coated calcium carbonate nodules) and PC (pedogenic $CaCO_3$; Pal et al. 2000, 2009a). Vertisols with Fe–Mn coated $CaCO_3$ are older than those with PCs (Mermut and Dasog 1986) that are formed in soils of dry climate soils (Pal et al. 2000). NPCs were formed in a much wetter climate than the present climate, ensuring adequate water for reduction and oxidation of iron and manganese to form Fe–Mn coatings. Although the Vertisols contain both silt and clay sized muscovite and biotite mica (Pal 2003), dioctahedral smectite (DOS) cannot be formed at the expense of muscovite (dioctahedral mica) because the weathering of muscovite is very

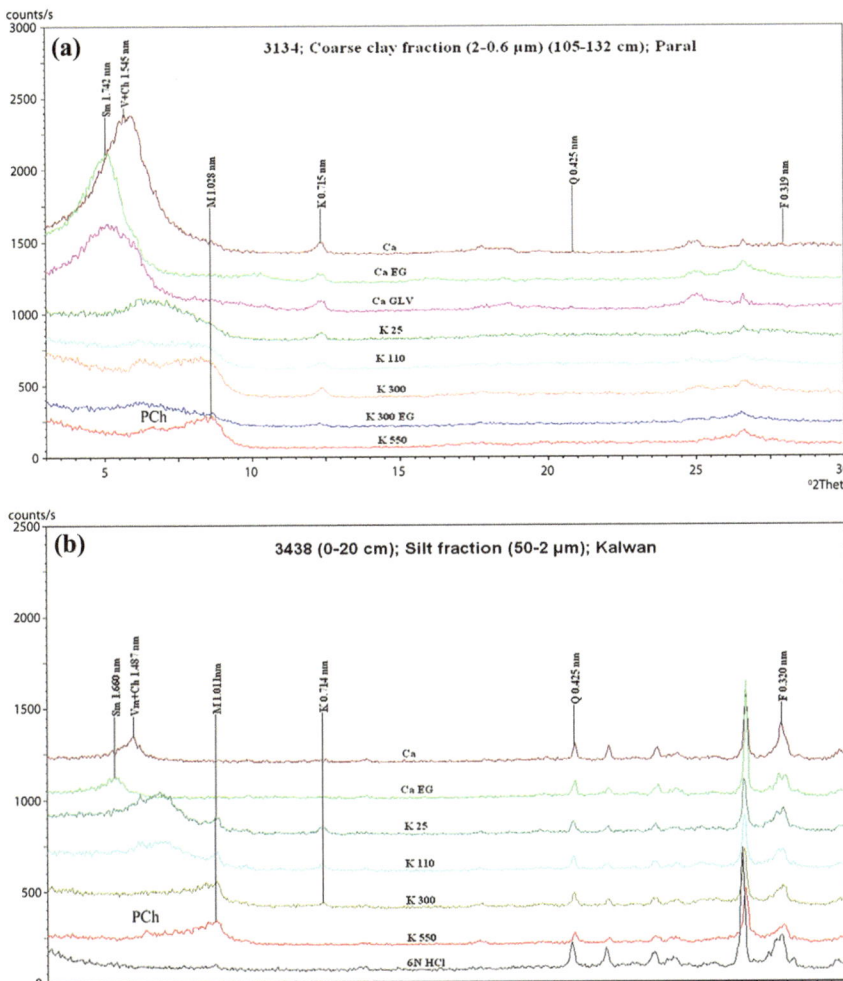

Fig. 2.9 Representative X-ray diffractograms of coarse clay (**a**), and silt (**b**) fractions of Vertisols of Peninsular India; Ca = Ca saturated; Ca-EG = Ca saturated plus ethylene glycol vapour treated; K25/110/300/550 °C = K-saturated and heated to 25, 110, 300, 550 °C; 6NHCl = 6N HCl treated silt fraction; Sm = Smectite, V + Ch = vermiculite plus chlorite; PCh = Pseudo chlorite; K = Kaolin; F = Feldspars; Q = Quartz (Adapted from Pal et al. 2003b; Bhople 2010)

sensitive to potassium levels in soils. Biotite converts to trioctahedral vermiculite (TOV; Pal 2003); thus, the simultaneous formation of DOS and TOV from mica is very unlikely (Pal et al. 1989; Ray et al. 2006a). Moreover, in the sub-humid and semi-arid climates that facilitate the formation of $CaCO_3$ from plagioclase (Pal et al. 2012b), mica may not yield as much DOS as required for Vertisols. The large quantity of DOS is, therefore, formed under a previous, humid climate regime in the source area as an alteration product of plagioclase (Pal et al. 1989; Srivastava et al. 1998). The formation of smectite from biotite is quite unlikely in a humid climate (Pal et al. 1989), but vermiculite could have transformed to HIV, which further transforms to PCh under acidic conditions. Hence, the formation of HIS did not continue for long in the HT climate, as evidenced from the presence of very small quantities of clay kaolin (Sm–K). In the event of prolonged weathering of HIS, kaolin would have become dominant mineral (Bhattacharyya et al. 1993). Thus, the HIS in the Vertisols were formed under a previous, more humid climate regime and its crystallinity, and also the HIV and PCh were preserved in the non-leaching environment of the latter sub-humid to dry climates (Pal et al. 2009a, 2012b).

2.5 Recent Advances in Pedogenic Processes in Vertisols

Although an extensive research on Vertisols of the Indian sub-continent was made in the past (Murthy et al. 1982; Murthy 1988), during the last decade and a half, the focus of Vertisol research has changed qualitatively (Pal et al. 2012a). Mineralogical, micro-morphological and age-control tools were used to measure the relatively subtle pedogenetic processes in Vertisols that have implications to their polygenesis (Pal et al. 2009a, b, c, 2012a, b, c), paleopedology (Pal et al. 2009a, b, c, 2012a, b, c) and edaphology (El-Swaify et al. 1985; Srivastava et al. 2002; Kadu et al. 2003; Bhattacharyya et al. 2009). Hence, to place recent research in the context of past research, critical appraisal of some important basic issues in Vertisol pedogenesis with regard to (a) importance of clay illuviation over pedoturbation, (b) relative rapidity of clay illuviation, pedoturbation and slickenside formation and developments of cracks, and (c) evolution sequences in the genesis of Vertisols is warranted.

2.5.1 Clay Illuviation in Vertisols: An Example of Proanisotropism

Earlier reviews of the past research work (Ahmad 1983; Murthy et al. 1982) explained that the distribution of clay is uniform throughout Vertisols because haploidisation within the pedon caused considerable pedoturbation (Mermut et al. 1996). But some studies reported that in selected cases, there is a gradual increase in clay content with depth (Dudal 1965); although, it was thought that the increase in clay content with depth is due to inheritance from parent material (Ahmad 1983).

Recent studies on Vertisols (that have no stratification in the parent material and no clay skins) indicated that the Bss horizons are substantially clay enriched with clay even up to ∼20%; an increase from the eluvial horizon (Pal et al. 2009a, 2012a). Micro-morphological investigation of the thin sections indicates the presence of >2% impure clay pedofeatures (Fig. 2.10a), which confirm that the clay is

Fig. 2.10 Representative photograph in cross polarised light. **a** Impure clay pedofeatures, **b** weakly oriented clay pedofeatures and **c** undifferentiated clay pedofeatures (Adapted from Pal et al. 2009a)

enriched in the Bss horizons of Vertisols by clay illuviation. Therefore, such Vertisols can also have argillic horizons (Pal et al. 2009a, 2012a), which justify the subsoil horizon designation as 'B' (Soil Survey Staff 2014) instead of 'A' in earlier concept on Vertisols with no horizonation (Soil Survey Staff 1975). The clay illuviation process in Indian Vertisols is no exception as it is observed in operation in clay soils with vertic properties in Canada (Dasog et al. 1987), Uruguay (Wilding and Tessier 1988) and Argentina (Blokhuis 1982). Pedoturbation was too much favoured as an important pedogenic process in Vertisols by the past researchers till early nineties (Soil Survey Staff 1992) that would obliterate all evidence of illuviation, except in the lower horizons (Eswaran et al. 1988; Mermut et al. 1996). The emphasis on pedoturbation possibly led Johnson et al. (1987) to consider this process to be an example of proisotropic pedoturbation caused by argilli-turbation, which was thought to destroy horizons or soil genetic layers and to make Vertisols revert to a simpler state. The recent evidence of clay-enriched Bss horizons caused by illuviation suggests that the argilli-turbation is not a primary pedogenetic process in simpler state. The clay-enrichment of Bss horizons via illuviation suggests that the argilli-turbation is not a primary pedogenetic process in Vertisols (Pal et al. 2009a) and represents a proanisotropism in the soil profile. This finding is reinforced by a steady decrease in soil organic carbon and by increases in $CaCO_3$, exchangeable magnesium percentage (EMP), exchangeable sodium percentage (ESP), water-dispersible clay (WDC) and carbonate clay (fine earth based) with depth (Table 2.4). Therefore, pedoturbation in Vertisols is a partially functional process that is not able to overshadow the more significant long-term clay illuviation process. Although argillic horizons are common in Vertisols, the Bt horizon does not get better than their dominant property (slickensides) because Vertisol soil order keys out before the Alfisols according to the US Soil Taxonomy. Thus, the classification of these soils as Vertisols would still be continued (Pal et al. 2009a, b, 2012a).

2.5.2   Factors of Clay Illuviation in Vertisols

The fine-sized smectite clay with a high surface area has all conditions required for dispersions, translocation and accumulation in subsurface horizons in Vertisols. This fact is evident from the considerable amount of WDC in majority of calcareous Vertisols, which increases with depth (Table 2.4). The presence of WDC in Vertisols indicate that the dispersion of clay smectite is possible under slightly acidic to moderately alkaline pH conditions at a very low electrolyte concentration (ECe $\leq$ 1 me L^{-1}; Table 2.4) that ensure a pH higher than the zero point of charge required for a full dispersion of clay (Eswaran and Sys 1979). Thus, for the illuviation of clay removal of carbonate is not pre-requisite as postulated by many researchers (Pal et al. 2003a), who thought Ca^{2+} ions enhance the flocculation and immobilization of colloidal material. Indian calcareous Vertisols have low quantities of soluble Ca^{2+} ions ($\ll$5me L^{-1}, Table 2.4) that are not sufficient to cause flocculation of clay particles. Therefore, movement of deflocculated fine clay smectite (and its subsequent accumulation in the Bss horizons) is possible in non-calcareous as well as calcareous Vertisols (Pal et al. 2012a). The primary source of Ca^{2+} ions in Vertisols' solution is the dissolution of NPCs (Srivastava et al. 2002). The depth distribution of EMP, ESP, carbonate clay, and soluble Na$^+$ ions in the majority of Vertisols in India (Table 2.4) suggests that the precipitation of CaCO$_3$ as PC enhances the pH and the relative abundance of Na$^+$ ions in soil exchange and in solution. The Na$^+$ ions in turn cause dispersion of clay smectites, and the dispersed smectites translocate even in the presence of CaCO$_3$. The formation of PC creates a chemical environment that facilitates the deflocculation of clay particles and their subsequent movement downward. Therefore, the PC formation and clay illuviation are two concurrent and contemporary pedogenic events that provide examples of pedogenic thresholds in dry climates (Pal et al. 2003a, 2009a, 2012a).

2.5.3   Relative Rapidity of Clay Illuviation, Pedoturbation and Slickenside Formation

Many researchers (Parsons et al. 1973; White 1967; Yaalon 1971) considered that the formation of slickensides is a very rapid pedogenic process as the Vertisols are

Table 2.4 Physical and chemical properties of Sodic Haplusterts[a] as representative of Vertisols of Peninsular India

(a) Physical properties

Lab. no	Horizon	Depth (cm)	Size class and particle diameter (mm) Total (% of < 2 mm)			Fine clay (%)	Fine clay/total clay (%)	BD Mg/m³	COLE	HC[b] cm/hr	WDC (%)
			Sand (2–0.05)	Silt (0.05–0.002)	Clay (< 0.002)						
3114	Ap	0–14	0.9	36.7	62.4	26.7	42.8	–	0.28	1.1	6.6
3115	Bw1	14–40	0.9	34.2	64.9	26.7	41.1	1.5	0.26	2.1	13.9
3116	Bw2	40–59	0.8	33.3	65.9	28.9	43.8	1.6	0.26	1.0	14.8
3117	Bss1	59–91	1.3	35.3	63.4	29.0	45.7	1.5	0.29	0.5	6.4
3118	Bss2	91–125	2.4	37.3	60.3	28.7	47.6	1.5	0.25	0.4	7.6
3119	Bss3	125–150	1.9	38.1	60.0	25.7	42.8	1.6	0.25	0.3	10.0

(b) Moisture at various tensions

Horizon	Depth (cm)	Moisture retention%							AWC
		33 kPa	100 kPa	300 kPa	500 kPa	800 kPa	1000 kPa	1500 kPa	
Ap	0–14	40.1	35.1	30.9	28.3	25.1	22.5	20.3	19.7
Bw1	14–40	41.7	37.3	30.6	28.2	26.2	25.1	19.1	22.7
Bw2	40–59	42.4	40.3	32.2	30.0	26.9	26.8	22.2	20.2
Bss1	59–91	43.9	43.1	33.2	32.6	28.5	27.9	19.8	24.1
Bss2	91–125	43.5	42.7	32.8	32.6	27.8	25.7	19.5	24.1
Bss3	125–150	48.5	42.7	37.5	33.0	29.4	28.3	26.2	22.3

(continued)

Table 2.4 (continued)

(c) Chemical properties

Depth (cm)	pH water (1:2)	CaCO$_3$ (%)	OC (%)	Extractable bases (cmol(p+)/kg^{-1})					CEC (cmol(p+)/kg^{-1})	Clay CEC (cmol (p+)/kg^{-1})	B.S. (%)
				Ca	Mg	Na	K	Sum			
0–14	7.8	9.3	0.81	46.2	14.4	0.6	1.0	62.2	65.2	99	95
14–40	7.9	9.4	0.66	43.4	15.6	2.1	0.7	61.0	61.8	94	98
40–59	8.0	10.7	0.59	42.0	17.8	2.7	0.7	63.0	63.5	95	99
59–91	8.4	11.0	0.61	38.2	20.2	4.2	0.7	63.3	63.5	100	99
91–125	8.5	13.7	0.48	28.9	22.0	5.8	0.6	57.3	62.2	95	92
125–150	8.5	15.6	0.42	25.8	22.4	8.6	1.1	57.9	66.7	96	87

(d) Exch. Ca/Mg, ECP, EMP, ESP and carbonate clay in soil and on fine earth basis (feb)

Depth (cm)	Exch. Ca/Mg	ECP	EMP	ESP	CO$_3$ clay (%)	CO$_3$ clay (feb) (%)
0–14	3.2	71	22	1.0	0.4	0.2
14–40	2.8	70	25	2.1	0.4	0.2
40–59	2.3	66	28	4.4	1.2	0.8
59–91	1.9	60	32	6.6	3.2	2.0
91–125	1.3	46	35	9.3	1.8	1.1
125–150	1.1	38	33	12.9	1.9	1.1

(e) Saturation extract analysis

Depth (cm)	Soluble cations (meq/l)							Soluble anions (meq/l)					RSC	SAR
	Sat.%	ECe	Ca	Mg	Na	K	Sum	CO$_3$	HCO$_3$	Cl	SO$_4$	Sum		
0–14	72.8	0.3	1.43	0.8	3.2	0.1	5.53	–	3.2	0.8	1.53	5.50	0.97	3.0
14–40	70.4	0.3	0.67	0.4	0.8	0.04	1.91	–	1.3	0.6	–	1.90	0.23	1.1
40–59	73.0	–	0.46	0.3	1.1	0.05	1.90	1.0	0.5	0.7	–	2.20	0.74	1.8
59–91	77.1	0.4	0.39	0.3	1.7	0.03	2.46	1.0	1.0	0.9	–	2.90	1.31	3.0
91–125	63.3	4.7	0.72	0.5	6.0	1.43	8.65	1.0	3.0	0.2	4.45	8.65	2.78	8.0
125–150	85.9	–	0.49	0.3	6.3	0.05	7.14	2.0	4.0	0.1	1.04	7.14	5.21	10.0

Adapted from Pal et al. (2003a)

[a]As defined by Pal et al. (2006a); [b]9 mm h^{-1} is the HC (WM) in 0–100 cm depth of soil

Fig. 2.11 Representative photograph of cross polarized light: poorly separated plasma in Vertic Haplustalfs (**a**), strong parallel plasmic fabric in Typic Haplusterts (**b**), and mosaic/stippled-speckled plasmic fabric in Aridic/Sodic Haplusterts (**c**) of Peninsular India (Adapted from Pal et al. 2009a, c)

formed on geomorphic surfaces that are <200–550 year old (Blokhuis 1982; Parsons et al. 1973). After their formation slickensides approach equilibrium with their environment in a period ranging from 100 to 1000 years (Yaalon 1971). A Vertic Haplustalf <100 year age (^{14}C age, Pal et al. 2006a) developed in the alluvium of the central Indian Deccan basalt during the SHD climate regime, exhibits pressure faces but lacks in slickensides and clay skins; however, it exhibits weakly oriented clay pedofeatures (Fig. 2.10b), undifferentiated clay pedofeatures (Fig. 2.10c) and poorly separated plasma (Fig. 2.11a). Such Vertic Haplustalfs have >8% more clay in the B horizons than in the Ap horizons, and the fine clay/total clay ratio in the B horizon is >1.2 times greater than that of the Ap horizon. Despite having vertic character, the thin sections of the soils did not show any of the disrupted clay pedofeatures that could be expected in soils with high COLE (>0.10, Pal et al. 2009a). The illuviation of clay in the absence of slickensides therefore indicates that illuviation is a faster pedogenetic process than the formation of slickensides, which possibly takes place within a 100-year span. The occurrence of shrink-swell soils over 500–1500 year (TL ages) with the illuviated clay features and slickensides of the eastern lower Indo-Gangetic Plains (IGP) under an SHM climate regime (Singh et al. 1998) suggests that a minimum time of 500 years is required to form slickensides in Vertisols. Therefore, intensive pedoturbation is not essential or important in the creation of typical morphogenetic characteristics in a Vertisol (Yaalon and Kalmar 1978).

2.5.4 Development of Microstructures and Vertical Cracks in Vertisols as Controlled by Smectite Swelling

Horizontal and vertical stresses in Vertisols are induced by abundance of smectite. In the upper horizons, low overburden pressure and cracks prevent the development of high lateral stresses. In the subsoils, where sphenoids and/or slickensides are formed, the difference between horizontal stress and vertical stress is quite large. These two sets of stresses act on soil during its swelling. As a result, failure occurs when the vertical stress is confined and the lateral stress exceeds the shear strength of the soils. Failure occurs along a grooved shear plane (theoretically 45° to the horizontal; Wilding and Tessier 1988). In reality, such shear failure may range from 10° to 60° (Knight 1980). The shear failure is manifested as the appearance of poro/parallel/reticulate/grano-striated plasmic fabric, indicating a prominent surface-oriented plasma separation (Fig. 2.11b) or stipple-speckled/mosaic-speckled/crystallitic plasmic fabric related to poor plasma separation in the Bss horizons (Fig. 2.11c; Kalbande et al. 1992; Pal et al. 2009a, b). The presence of sphenoids and/or slickensides and the dominance of poro/parallel/grano/reticulate-striated plasmic fabric in Indian Vertisols in HT and SHM climates indicate that the shrink–swell activity of smectites has been extensive. In contrast, the dominance of stippled/mosaic-speckled plasma in SHD soils, mosaic/crystallitic plasma in SAM soils, mosaic/stippled-speckled plasma in SAD soils and crystallitic plasma in AD soils clearly suggests that shrink–swell is much less significant in the soils of drier climates compared to HT and SHM climates, and it manifests in poor plasma separation. It is, therefore, understood that weak swelling of smectite is sufficient for the development of sphenoids and/or slickensides, but it is not definitely adequate to cause strong plasma separation (Pal et al. 2001, 2009a, b), even when the soils have almost identical COLE values and comparable amounts of expansible clays (Pal et al. 2006a, 2009a). The swelling of smectites in these soils, however, has been restricted neither by the presence of $CaCO_3$ and calcite crystals, nor by the decrease in smectite interlayer surface area by partial hydroxy-interlayering (Pal et al. 2009a).

It is interesting to note that the saturated hydraulic conductivity (sHC) of all Vertisols is not identical but decreases rapidly with depth. However, the decrease is sharper in SAD and AD soils because of their subsoil sodicity (ESP > 5, Pal et al. 2009a). The reduced sHC restricts the vertical and lateral movement of water in the subsoils. As a consequence, during the very hot summer months (April–June), subsoils of SAD and AD Vertisols would have less water. This deficit is manifested in the form of the deep cracks cutting through their Bss horizons, in contrast to higher MAR soils in which the cracks do not extend beyond the slickensided horizon at 40–50 cm (Fig. 2.2). Thus, the lack of adequate soil water during the shrink–swell cycles restricts the swelling of smectite and results in weaker plasma separation in SAD and AD soils (Pal et al. 2009a, b). The SAM, SAD and AD subsoils remain, in general, under less amount of water compared to those of HT, SHM, and SHD climates during the Holocene period. Such dryness causes the modifications in subsoils of Vertisols in terms of subsoil sodicity, poor plasma separation, and cracks cutting through the Bss horizons due to the accelerated formation of PC. Therefore, such Vertisols qualify to be polygenetic soils (Pal et al. 2001, 2009a, 2012a).

2.6 Evolutionary Pathways of Vertisol Formation

The U.S. Soil Taxonomy recognizes 'intergrade' between Vertisols and soils of other orders with vertic character. Thus, successive stages of pedogenic evolution in Vertisols were conceptualized by Blokhuis (1982), who thought that Vertisols would lose their vertic characters and subsequently convert to non-vertic soils.

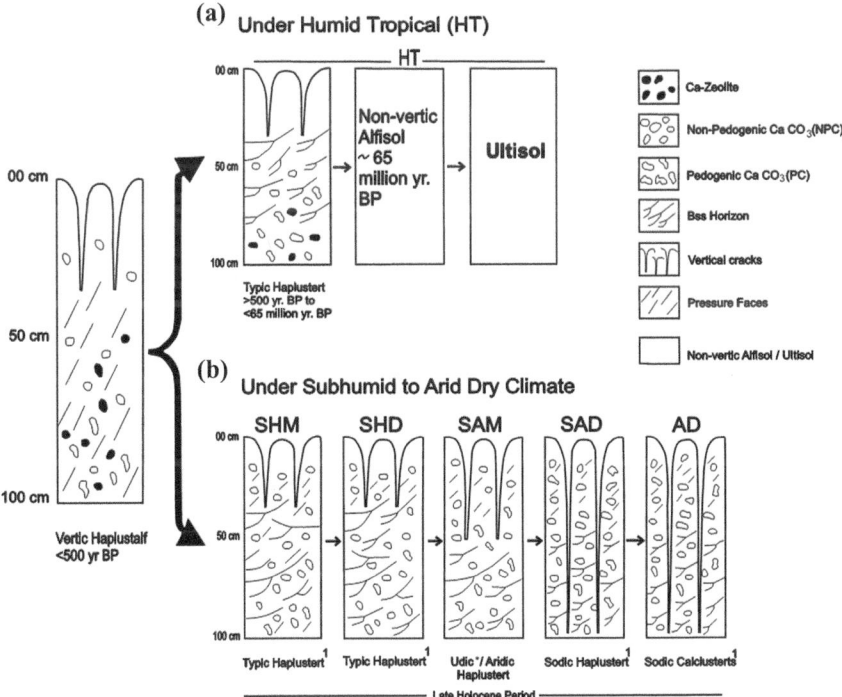

Fig. 2.12 Successive pathways of pedogenic evolution in Vertisols (Adapted from Pal et al. 2009b)

Eswaran et al. (1988), on the other hand, suggested that a Vertisol (pH < 6.5) in surface horizons would form an argillic horizon as leaching advanced, and with the accumulation of translocated clay, the soils may qualify as Vertic Haplustalfs. In these conceptualizations the presence of argillic horizon in original Vertisols was not envisaged.

Studies on the evolution of soils in HT parts of the Western Ghats (Bhattacharyya et al. 1993, 1999) and north-eastern (Bhattacharyya et al. 2000), and southern India (Chandran et al. 2005) suggest that with time, Vertisols in the presence of Ca-zeolites (Typic Haplusterts) of HT remain as Vertisols as long as the zeolites continue to provide bases, which prevents the total transformation of smectites to kaolin (Fig. 2.12a). However, when the stocks of zeolites are depleted completely, the soils would gradually become acidic and kaolinitic and phase towards Ultisols through an intermediate stage of non-vertic Alfisols (Fig. 2.12a). Silica is insoluble in an acidic environment; therefore, the complete transformation of smectite to kaolinite is improbable. The end result will be that Ultisols would remain unchanged, with Sm–K as the dominant minerals (Chandran et al. 2005; Pal et al. 2014).

A recent extensive pedogenetic study of Indian Vertisols in a climosequence expands the basic understanding of Vertisol evolution from Typic Haplusterts to Udic/Aridic/Sodic Haplusterts and Sodic Calciusterts (Pal et al. 2009a; Fig. 2.12b). These Vertisols may remain in equilibrium with their climatic environments until the climate changes further, after which another pedogenic threshold is reached. These soils are of Holocene period but exit as the products of polygenic evolution. Due to subsoil sodicity caused by illuviation of Na-clay smectites in Vertisols of the

Fig. 2.13 Progressive development of sodicity due to illuviation of Na-clay in Vertisols while aridity continues with time (Adapted from Pal et al. 2012b). *ESP* exchangeable sodium percentage

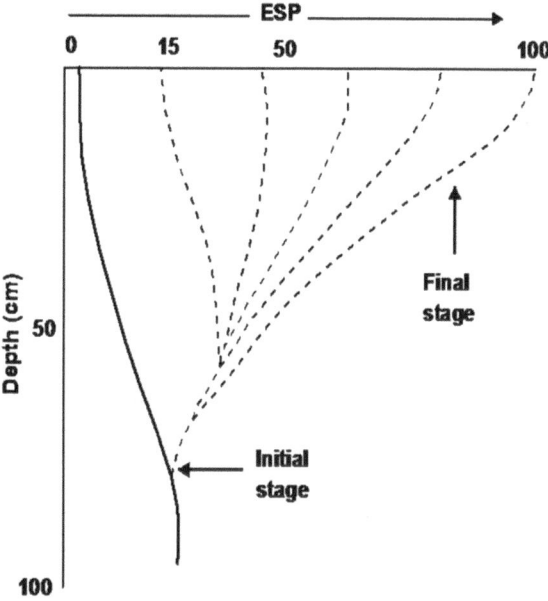

SAM, SAD and AD climates, the initial impairment of the percolative moisture regime would create a soil system in which gains exceed losses. This self-terminating process (Yaalon 1971) would lead to the development of sodic soils in which ESP finally decreases with depth if aridity continues (Fig. 2.13), and the formation of such sodic soils exhibit regressive pedogenesis (Johnson and Watson-Stegner 1987; Pal et al. 2013b).

References

Ahmad N (1983) Vertisols, pedogenesis and soil taxonomy. In: Wilding LP, Smeck NE, Hall GF (eds) The soil orders, vol. II. Elsevier, Amsterdam, pp 91–123

Ahmad N (1996) Occurrence and distribution of Vertisols. In: Ahmad N, Mermut A (eds) Vertisols and technologies for their management. Elsevier, Amsterdam, pp 1–41

Alexiades CA, Jackson ML (1965) Quantitative determination of vermiculite in soils. Soil Sci Soc Am Proc 29:522–527

Bardaoui M, Bloom PR (1990) Iron-rich high-charge Beidellite in vertisols and mollisols of the high Chaouia region of Morocco. Soil Sci Soc Am J 54:267–274

Barnhisel RI, Bertsch PM (1989) Chlorites and hydroxy-interlayered vermiculites and smectite, minerals in soil environments, second edition. In: Dixon JB, Weed SB (eds) Soil science society of America book series (Number 1). Wisconsin, USA, pp 729–788

Beckman GG, Thompson CH, Hubble GD (1974) Genesis of red and black soils on basalt on the Darling Downs, Queensland, Australia. J Soil Sci 25:215–280

Bhattacharyya T, Pal DK, Deshpande SB (1993) Genesis and transformation of minerals in the formation of red (Alfisols) and black (Inceptisols and Vertisols) soils on Deccan Basalt in the Western Ghats, India. J Soil Sci 44:159–171

Bhattacharyya T, Pal DK, Deshpande SB (1997) On kaolinitic and mixed mineralogy classes of shrink–swell soils. Aust J Soil Res 35:1245–1252

Bhattacharyya T, Pal DK, Srivastava P (1999) Role of zeolites in persistence of high altitude ferruginous Alfisols of the humid tropical Western Ghats, India. Geoderma 90:263–276

Bhattacharyya T, Pal DK, Srivastava P (2000) Formation of gibbsite in presence of 2:1 minerals: an example from Ultisols of northeast India. Clay Miner 35:827–840

Bhattacharyya T, Pal DK, Chandran P, Ray SK (2005) Land-use, clay mineral type and organic carbon content in two Mollisols–Alfisols–Vertisols catenary-sequences of tropical India. Clay Res 24:105–122

Bhattacharyya T, Ray SK, Pal DK, Chandran P (2009) Mineralogy class of calcareous zeolitised Vertisols. Clay Res 28:73–82

Bhattacharyya et al (2014) Georeferenced soil information system: assessment of database. Curr Sci 107:1400–1419

Bhople BS (2010) Layer charge characteristics of some Vertisol clays of Maharashtra and its relationship with soil properties and management. Ph. D Thesis, Dr. P D K V, Akola, Maharashtra, India

Bhople BS, Pal DK, Ray SK, Bhattacharyya T, Chandran P (2011) Seat of charge in clay smectites of some Vertisols of Maharashtra. Clay Res 30:15–27

Blokhuis AA (1982) Morphology and genesis of Vertisols. In: Vertisols and rice soils in the tropics. Transactions 12th international congress of soil science, New Delhi, pp 23–47

Borchardt G (1989) Smectites. In: Dixon JB, Weed SB (eds) Minerals in soil environments. Soil Science Society of America, Madison, Wisconsin, pp 675–727

Buol SW, Hole FD, McCracken RJ (1978) Soil genesis and classification. Oxford and IBH Publ. Co., New Delhi, India

Chandran P, Ray SK, Bhattacharyya T, Srivastava P, Krishnan P, Pal DK (2005) Lateritic soils of Kerala, India: their mineralogy, genesis and taxonomy. Aust J Soil Res 43:839–852

Chen CC, Turner FT, Dixon JB (1989) Ammonium fixation by high charge smectite in selected Texas Gulf Coast soils. Soil Sci Soc Am J 53:1035–1040

Coulombe CE, Wilding LP, Dixon JB (1996) Overview of Vertisols: characteristics and impacts on society. In: Sparks DL (ed) Adv Agron, vol 57. Academic Press, New York, pp 289–375

Dasog GS, Acton DF, Mermut AR (1987) Genesis and classification of clay soils with vertic properties in Saskatchewan. Soil Sci Soc Am J 51:1243–1250

Deshmukh HV, Chandran P, Pal DK, Ray SK, Bhattacharyya T, Potdar SS (2014) A pragmatic method to estimate plant available water capacity (PAWC) of rainfed cracking clay soils (Vertisols) of Maharashtra, Central India. Clay Res 33:1–14

Dudal R(1965) Dark clay soils of tropical and subtropical regions. FAO Agric. Dev. Paper, 83. FAO, Rome. 161 pp

Dudal R, Eswaran E (1988) Distribution, properties and classification of Vertisols. In: Wilding LP, Puentes R (eds) Vertisols: their distribution, properties, classification and management. Texas A&M University Printing Centre, College Station, Texas, pp 1–22

Duncan RA, Pyle DG (1988) Rapid eruption of the Deccan flood basalt, western India, In: Subbarao KV (ed) Deccan flood basalts. Geological Society of India Memoir No. 10, Geological Society of India, Bangalore, pp 1–19

El-Swaify SA, Pathak P, Rego TJ, Singh S (1985) Soil management for optimized productivity under rain-fed conditions in the semi-arid tropics. Adv Soil Sci 1:1–63

Eswaran H, Sys C (1979) Argillic horizon in LAC soils formation and significance to classification. Pedologie 29:175–190

Eswaran H, Kimble J, Cook T (1988) Properties, genesis and classification of Vertisols. In: Hirekerur LR, Pal DK, Sehgal JL, Deshpande SB (eds) Classification, management and use potential of swell–shrink soils. INWOSS, 24–28 Oct 1988. National Bureau of Soil Survey and Land Use Planning, Nagpur, pp 1–22

Ghosh SK, Kapoor BS (1982) Clay minerals in Indian soils. In: Review of soil research in India. Transactions of the 12th international congress of soil science, New Delhi, India. pp 703–710

Greene-Kelly R (1953) The identification of montmorillonoids in clays. J Soil Sci 4:233–237

Hajek BF (1985) Mineralogy of Aridisols and Vertisols. Proceedings 5th international soil classification. Soil Survey Administration, Khartoum, Sudan, pp 221–230

Harward ME, Carstea DD, Sayegh AH (1969) Properties of vermiculites and smectites: expansion and collapse. Clays Clay Miner 16:437–447

Heidari A, Mahmoodi Sh, Roozitalab MH, Mermut AR (2008) Diversity of clay minerals in the Vertisols of three different climatic regions in western Iran. J Agric Sci Technol 10:269–284

Hofmann U, Klemen R (1950) Verlüst der Austaüschfahigkeit von Litiümionen and Bentonite dürch Erhitzüng. Zeitschrift für Inorganische Chemie 262:95–99

Johnson DL, Watson-Stegner D (1987) Evolution model of pedogenesis. Soil Sci 143:349–366

Johnson DL, Watson-Stegner D, Johnson DN, Schaetzl RJ (1987) Proisotropic and proanisotropic processes of pedoturbation. Soil Sci 143:278–292

Kadu PR (1997) Soils of Adasa Watershed: their geomorphology, formation, characterisation and land evaluation for rational land use. Ph. D. thesis, Dr. P D K V, Akola, Maharashtra, India

Kadu PR, Vaidya PH, Balpande SS, Satyavathi PLA, Pal DK (2003) Use of hydraulic conductivity to evaluate the suitability of Vertisols for deep-rooted crops in semi-arid parts of central India. Soil Use Manag 19:208–216

Kalbande AR, Pal DK, Deshpande SB (1992) b-fabric of some benchmark Vertisols of India in relation to their mineralogy. J Soil Sci 43:375–385

Kapse VK, Ray SK, Chandran P, Bhattacharyya T, Pal DK (2010) Calculating charge density of clays: an improvised method. Clay Res 29:1–13

Knight MJ (1980) Structural analysis and mechanical origins of gilgai at Boorook, Victoria, Australia. Geoderma 23:245–283

Lagaly G (1994) Layer charge determination by alkyl ammonium ions. In: Mermut AR (ed) Layer charge characteristics of 2:1 silicate clay minerals. CMS workshop lectures, vol. 6, The Clay Minerals Society, Boulder, USA. pp 1–46

Laird DA, Fenton TE, Scott AD (1988) Layer charge of smectites in an Argialboll-Argiaquoll sequence. Soil Sci Soc Am J 52:463–467

Malla PB, Douglas LA (1987) Identification of expanding layer silicates: layer charge vs expansion properties. In: Schultz LG, Van Olphen H, Mumpton FA (eds) Proceedings of the international clay conference. Denver, Clay Mineral Society, Bloomington, Indiana, USA, pp 277–283

Mandal et al (2014) Revisiting agro-ecological sub-regions of India—a case study of two major food production zones. Curr Sci 107:1519–1536

Mermut AR, Dasog GS (1986) Nature and micromorphology of carbonate glaebules in some Vertisols of India. Soil Sci Soc Am J 50:382–391

Mermut AR, Padmanabham E, Eswaran H, Dasog GS (1996) Pedogenesis. In: Ahmad N, Mermut AR (eds) Vertisols and technologies for their management. Elsevier, Amsterdam, pp 43–61

Mohr ECJ, Van Baren FA, Van Schuylenborgh J (1972) Tropical soils—a comprehensive study of their genesis. Mouton-Ichtiarbaru-Van Hoeve, The Hague

Murthy ASP (1988) Distribution, properties and management of Vertisols of India. Adv Soil Sci 8:151–214

Murthy RS, Bhattacharjee JC, Landey RJ, Pofali RM (1982) Distribution, characteristics and classification of Vertisols. In: Vertisols and rice soils of the tropics, Symposia paper II, 12th international congress of soil science, New Delhi. Indian Society of Soil Science, pp 3–22

Myers RJK, Pathak P (2001) Indian Vertisols: ICRISAT's research impact-past, present and future. In: Syers JK, Penning de Vries, FT, Nyamudeza P (eds) The sustainable management of Vertisols. CAB International Publishing, Wallingford, pp 203–219

NBSS&LUP (2002) Soils of India, NBSS Pub. 94. NBSS and LUP, Nagpur, India

Pal DK (1985) Potassium release from muscovite and biotite under alkaline conditions. Pedologie (Ghent) 35:133–146

Pal DK (1988) On the formation of red and black soils in southern India. In: Hirekerur LR, Pal DK, Sehgal JL, Deshpande SB (eds) Transactions international workshop swell-shrink soils. Oxford and IBH, New Delhi, pp 81–82

Pal DK (2003) Significance of clays, clay and other minerals in the formation and management of Indian soils. J Indian Soc Soil Sci 51:338–364

Pal DK (2008) Soils and their mineral formation as tools in paleopedological and geomorphological studies. J Indian Soc Soil Sci 56:378–387

Pal DK, Deshpande SB (1987) Characteristics and genesis of minerals in some benchmark Vertisols of India. Pedologie (Ghent) 37:259–275

Pal DK, Durge SL (1987) Potassium release and fixation reactions in some benchmark Vertisols of India in relation to their mineralogy. Pedologie 37:103–116

Pal DK, Deshpande SB, Venugopal KR, Kalbande AR (1989) Formation of di- and trioctahedral smectite as an evidence for paleoclimatic changes in southern and central Peninsular India. Geoderma 45:175–184

Pal DK, Dasog GS, Vadivelu S, Ahuja RL, Bhattacharyya T (2000) Secondary calcium carbonate in soils of arid and semi-arid regions of India. In: Lal R, Kimble JM, Eswaran H, Stewart BA (eds) Global climate change and pedogenic carbonates. Lewis Publishers, Boca Raton, Florida, pp 149–185

Pal DK, Balpande SS, Srivastava P (2001) Polygenetic Vertisols of the Purna Valley of Central India. Catena 43:231–249

Pal DK, Srivastava P, Bhattacharyya T (2003a) Clay illuviation in calcareous soils of the semi-arid part of the Indo-Gangetic Plains, India. Geoderma 115:177–192

Pal DK, Bhattacharyya T, Ray SK, Bhuse SR (2003b) Developing a model on the formation and resilience of naturally degraded black soils of the peninsular India as a decision support system for better land use planning. NRDMS, Department of Science and Technology (Govt. of India) Project report, NBSSLUP (ICAR), Nagpur. 144 pp

Pal DK, Bhattacharyya T, Ray SK, Chandran P, Srivastava P, Durge SL, Bhuse SR (2006a)
    Significance of soil modifiers (Ca-zeolites and gypsum) in naturally degraded Vertisols of the
    peninsular India in redefining the sodic soils. Geoderma 136:210–228
Pal DK, Bhattacharyya T, Chandran P, Ray SK, Satyavathi PLA, Raja P, Maurya UK,
    Paranjape MV (2006b) Pedogenetic processes in a shrink–swell soil of central India.
    Agropedology 16:12–20
Pal DK, Nimkar AM, Ray SK, Bhattacharyya T, Chandran, P (2006c) Characterisation and
    quantification of micas and smectites in potassium management of shrink–swell soils in
    Deccan basalt area. In: Benbi DK, Brar MS, Bansal SK (eds) Balanced fertilization for
    sustaining crop productivity. Proceedings of the international symposium held at PAU,
    Ludhiana, India, 22–25 Nov 2006 IPI, Switzerland, pp 81–93
Pal DK, Dasog GS, Bhattacharyya T (2009a) Pedogenetic processes in cracking clay soils
    (Vertisols) in tropical environments of India: a critique. J Indian Soc Soil Sci 57:422–432
Pal DK, Bhattacharyya T, Chandran P, Ray SK (2009b) Tectonics-climate-linked natural soil
    degradation and its impact in rainfed agriculture: Indian experience. In: Wani SP, Rockström J,
    Oweis T (eds) Rainfed agriculture: unlocking the potential. CABI International, Oxfordshire,
    U.K., pp 54–72
Pal DK, Bhattacharyya, T, Chandran, P. Ray SK, Satyavathi PLA, Durge SL, Raja P, Maurya UK
    (2009c) Vertisols (cracking clay soils) in a climosequence of Peninsular India: evidence for
    Holocene climate changes. Quat Int 209:6–21
Pal DK, Sohan Lal, Bhattacharyya T, Chandran P, Ray SK, Satyavathi PLA, Raja P, Maurya UK,
    Durge SL, Kamble GK (2010) Pedogenic thresholds in benchmark soils under rice-wheat
    cropping system in a climosequence of the Indo-Gangetic alluvial plains. Nagpur, Final Project
    Report, Division of Soil Resource Studies, NBSS&LUP (ICAR), p 193
Pal DK, Wani SP, Sahrawat KL (2012a) Vertisols of tropical Indian environments: Pedology and
    edaphology. Geoderma 189–190:28–49
Pal DK, Bhattacharyya T, Wani SP (2012b) Formation and management of cracking clay soils
    (Vertisols) to enhance crop productivity: Indian experience. In: Lal R, Stewart BA(eds) World
    soil resources, Francis and Taylor, pp 317–343
Pal DK, Bhattacharyya T, Sinha R, Srivastava P, Dasgupta AS, Chandran P, Ray SK, Nimje A
    (2012c) Clay minerals record from late quaternary drill cores of the Ganga plains and their
    implications for provenance and climate change in the Himalayan Foreland. Palaeogeogr
    Palaeoclimatol Palaeoecol 356–357:27–37
Pal DK, Wani SP, Sahrawat KL (2013a) Zeolitic soils of the Deccan basalt areas in India: their
    pedology and edaphology. Curr Sci 105:309–318
Pal DK, Sarkar D, Bhattacharyya T, Datta SC, Chandran P, Ray SK (2013b) Impact of climate
    change in soils of semi-arid tropics (SAT). In: Bhattacharyya et al. (eds) Climate change and
    agriculture, Studium Press, New Delhi, pp 113–121
Pal DK, Wani SP, Sahrawat KL (2014) Srivastava P (2014) Red ferruginous soils of tropical
    Indian environments: a review of the pedogenic processes and its implications for edaphology.
    Catena 121:260–278. doi:10.1016/j.catena.05.023
Parsons RB, Moncharoan L, Knox EG (1973) Geomorphic occurrence of Pelloxererts, Wilamette
    Valley, Oregon. Soil Sci Soc Am Proc 37:924–927
Pillai M, Pal DK, Deshpande SB (1996) Distribution of clay minerals and their genesis in
    ferruginous and black soils occurring in close proximity on Deccan basalt plateau of Nagpur
    district, Maharashtra. J Indian Soc Soil Sci 44:500–507
Probert ME, Fergus IF, Bridge BJ, McGarry D, Thompson CH, Russel JS (1987) The properties
    and management of Vertisols. CAB International, Wallingford, Oxon, UK
Puentes R, Harris BL, Victoria C (1988) Management of Vertisols in temperate regions. In:
    Wilding LP, Puentes P (eds) Vertisols: their distribution, properties, classification and
    management. Texas A&M University Printing Center, College Station, Texas, pp 129–145
Ray SK, Chandran P, Bhattacharyya T, Durge SL (2002) Modification of Vertisol properties in
    relation to charge characteristics of smectites. Abstract, National Seminar, Indian society of soil
    science and 67th annual convention, Jabalpur, M.P., India, pp 227–228

Ray SK, Chandran P, Bhattacharyya T, Durge SL, Pal DK (2003) Layer charge of two benchmark Vertisol clays by alkylammonium method. Clay Res 22:13–27

Ray SK, Chandran P, Bhattacharyya T, Pal DK (2006b) Determination of layer charge of soil smectites by alkyl ammonium method: effect of removal of hydroxy-interlayering. Abstract, 15th annual convention and national symposium on "Clay Research in relation to agriculture, environment and industry" of the clay minerals society of India. BCKVV, Mohanpur, West Bengal, p 3

Ray SK, Bhattacharyya T, Chandran P, Sahoo AK, Sarkar D, Durge SL, Raja P, Maurya UK, Pal DK (2006a) On the formation of cracking clay soils (Vertisols) in West Bengal. Clay Res 25:141–152

Rich CI (1968) Hydroxy-interlayering in expansible layer silicates. Clays Clay Miner 16:15–30

Sawhney BL, Jackson ML (1958) Soil montmorillonite formulas. Soil Sci Soc Am Proc 22:115–118

Shirsath SK, Bhattacharyya T, Pal DK (2000) Minimum threshold value of smectite for vertic properties. Aust J Soil Res 38:189–201

Singh LP, Parkash B, Singhvi AK (1998) Evolution of the lower Gangetic Plain landforms and soils in West Bengal, India. Catena 33:75–104

Smeck NE, Runge ECA, Mackintosh EE (1983) Dynamics and genetic modeling of soil system. In: Wilding LP, Smeck NE, Hall GF (eds) Pedogenesis and soil taxonomy: I Concepts and interactions. Elsevier, New York, pp 51–58

Soil Survey Staff (1975) Soil taxonomy: a basic system of soil classification for making and interpreting soil surveys. Agriculture handbook no. 436, Soil Conservation Service, US Dept. of Agriculture

Soil Survey Staff (1992) Keys to soil taxonomy, fifth ed. SMSS technical monograph, no. 19. SMSS, Blacksburg, Virginia

Soil Survey Staff (1994) Keys to soil taxonomy, sixth ed. SMSS Technical monograph, no. 19. SMSS, Blacksburg, Virginia

Soil Survey Staff (1998) Keys to soil taxonomy, eight edn. USDA, NRCS, Washington, DC

Soil Survey Staff (1999) Soil taxonomy: a basic system of soil classification for making and interpreting soil surveys, USDA-SCS agricultural handbook no 436, 2nd edn. U.S. Govt, Printing Office, Washington, DC

Soil Survey Staff (2014) Keys to soil taxonomy (twelfth edition), United States Dept. of Agriculture, Natural Resource Conservation Service, Washington DC

Srivastava P, Parkash B, Pal DK (1998) Clay minerals in soils as evidence of Holocene climatic change, central Indo-Gangetic Plains, north-central India. Quatern Res 50:230–239

Srivastava P, Bhattacharyya T, Pal DK (2002) Significance of the formation of calcium carbonate minerals in the pedogenesis and management of cracking clay soils (Vertisols) of India. Clays Clay Miner 50:111–126

Srivastava P, Rajak M, Sinha R, Pal DK, Bhattacharyya T (2010) A high resolution micromorphological record of the late quaternary Paleosols from Ganga-Yamuna Interfluve: stratigraphic and Paleoclimatic implications. Quatern Int 227:127–142

Srivastava P, Pal DK, Aruche KM, Wani SP, Sahrawat KL (2015) Soils of the Indo-Gangetic Plains: a pedogenic response to landscape stability, climatic variability and anthropogenic activity during the Holocene. Earth Sci Rev 140: 54–71. doi:10.1016/j.earscirev.2014.10.010

Syers JK, Penning de Vries FT, Nyamudeza P (eds) (2001) The sustainable management of Vertisols. CAB International Publishing, Wallingford

Tessier D, Pedro G (1987) Mineralogical characterization of 2:1 clays in soils: importance of the clay texture. In: Schultz LG, van Olphen H, Mumpton FA (eds) Proceedings international clay conference, The Clay Minerals Society. Bloomington, IN, Denver, pp 78–84

Thakare PV, Ray SK, Chandran P, Bhattacharyya T, Pal DK (2013) Does sodicity in Vertisols affect the layer charge of smectites? Clay Res 32:76–90

Vaidya PH, Pal DK (2002) Micro topography as a factor in the degradation of Vertisols in central India. Land Degrad Dev 13:429–445

Weiss A, Koch G, Hofmann UK (1955) Zur kenntnis von Saponit. Ber dtsch. Keram. Ges. 32:12–17

White EM (1967) Soil age and texture factors in subsoil structure genesis. Soil Sci 103:288–298

Wilding LP, Coulombe CE (1996) Expansive soils: distribution, morphology and genesis. In: Baveye P, McBride MB (eds) Proceedings NATO-ARW on clay swelling and expansive soils. Kluwer Academic, Dordrecht, The Netherlands

Wilding LP, Tessier D (1988) Genesis of Vertisols: shrink–swell phenomenon. In: Wilding LP, Puentes R (eds) Vertisols: their distribution, properties, classification and management. Texas A&M University Printing Centre, College Station, Texas, pp 55–79

Wildman WE, Whittig LD, Jackson ML (1968) Serpentine stability in relation to formation of iron-rich montmorillonite in some California soils. Am Mineral 56:587–602

Wilson MJ (1987) X-ray powder diffraction methods. In: Wilson MJ (ed) A handbook of determinative methods in clay mineralogy. Chapman & Hall, New York, pp 26–98

Yaalon DH (1971) Soil forming processes in time and space. In: Yaalon DH (ed) Paleopedology. Israel University Press, Jerusalem

Yaalon DH, Kalmar D (1978) Dynamics of cracking and swelling clay soils: displacement of skeletal grains, optimum depth of slickensides and rate of intra-pedonic turbation. Earth Surf Proc 3:31–42

Yerima BPK, Calhoun FG, Senkayi AL, Dixon JB (1985) Occurrence of interstratified kaolinite–smectite in El Salvador Vertisols. Soil Sci Soc Am J 49:462–466

Yerima BPK, Wilding LP, Calhoun FG, Hallmark CT (1987) Volcanic ash-influenced Vertisols and associated Mollisols of El Salvador: physical, chemical and morphological properties. Soil Sci Soc Am J 51:699–708

Yousif AA, Mohamed HHA, Ericson T (1988) Clay and iron minerals in soils of the clay plains of central Sudan. J Soil Sci 39: 539–548

Chapter 3
Red Ferruginous Soils: Pedology, Mineralogy and Taxonomy

Abstract Red ferruginous (RF) soils of Indian tropical environments belong to five taxonomic soil orders (Entisols, Inceptisols, Alfisols, Mollisols and Ultisols), which clearly indicate that tropical RF soils in India have captured wide soil diversity. The spatially associated Ultisols with acidic Alfisols and Mollisols in both zeolitic and non-zeolitic parent materials in humid tropical (HT) climatic environments provides a unique example of tropical soil formation. Such soil formation discounts the exiting conceptual models on tropical soils. For a long time, this fact was not much appreciated, until the role of zeolites and other base rich parent materials was implicated in pedology and edaphology by the Indian soil scientists and soil mineralogists during the last two decades. Indian tropical soils support multiple production systems and generally maintain positive organic carbon balance. The recent developments on the pedology of RF soils, including their physical, chemical, biological, mineralogical and micro-morphological properties are very timely as the new knowledge improves the understanding as to how the parent material composition influences the formation of Alfisols, Mollisols and Ultisols in weathering environments of HT climate. This knowledge also explains how the relict Alfisols of semi-arid tropical (SAT) areas is polygenetic created by climate shift during the Holocene. Pioneering research efforts have improved the basic understanding of why the formation of Oxisols from Ultisols is an improbable genetic pathway in tropical environment of India and elsewhere in the world. There is a strong need to modify the mineralogy class of highly weathered RF soils. This basic information will help to dispel some of the myths on the formation of tropical soils and their low fertility.

Keywords Red ferruginous soils · Vertisols · Alfisols · Mollisols · Ultisols

© Springer International Publishing AG 2017
D.K. Pal, *A Treatise of Indian and Tropical Soils*,
DOI 10.1007/978-3-319-49439-5_3

3.1 Introduction

One of the main constraints to follow tropical soil science in its real perspective is the confusing and inaccurate terminology that existed throughout the literature (Sanchez 1976). Even in southern India, where pioneering work on laterites had started, soils previously considered to be 'laterite soils' are now classed as Alfisols, Inceptisols or Ultisols (Gowaikar 1973; Rengasamy et al. 1978). In the Indian sub-continent tropical soils include group of soils variously termed 'red', 'brown', 'yellow', 'laterite', 'lateritic', 'ferralitic' and 'latosols'. These soils lack precise definition, and therefore, they are termed as red ferruginous (RF) soils (Rengasamy et al. 1978). They are rubefied soils (7.5 YR to 2.5 YR hue) of tropical India characterized by reddish colour, stable structure mainly by the influence of iron oxides, poor base status, and highly to moderately acidic in reaction (Pal et al. 2000a). These soils mostly occur in the semi-arid to humid tropical and sub-tropical regions in eight agro-ecological regions of India. They cover one-fourth of the total geographical area of 329 million ha and are found in the states/union territories of Kerala, Tamil Nadu, Karnataka, Andhra Pradesh, Madhya Pradesh, Gujarat, Maharashtra, Goa, Orissa, Bihar, West Bengal, Sikkim, Manipur, Nagaland, Mizoram, Meghalaya, Arunachal Pradesh, Assam, Tripura, Pondicherry and Andaman and Nicobar islands (Sehgal et al. 1998; Bhattacharyya et al. 2009). These soils are included in the tropical soils of the world. Tropical soils comprise approximately 40% of the land surface of the earth, indicating that more than one-third of the soils of the world are tropical soils (Eswaran et al. 1992a). It could be, therefore, envisaged that research results developed on the agricultural management practices in the Indian subcontinent for enhancing crop productivity and maintaining soil health might also be applicable to similar soils elsewhere.

The Indian tropical soils support multiple production systems. Current information on their agricultural land uses (edaphology) demonstrates that they capture a wide variability in use for cereal production in the semi-arid, sub-humid, per-humid and coastal climatic environments. These soils also support forestry, horticultural, spices and cash crops, and thus contribute substantially to India's growing self-sufficiency in food production and food stocks. In general, the management of the Indian tropical RF soils is site-specific and requires an understanding of the degradation and regeneration processes to optimize their use by developing appropriate management practices (Sehgal 1998). Basic pedological research is needed to understand some of the unresolved edaphological issues of these soils to develop improved management practices. Thus, a critical appraisal of the Indian tropical RF soils and their pedogenetic processes that are linked to selected soil bulk properties will always remain a responsibility of the world soil scientists (Pal et al. 2014).

A basic inventory on the occurrence of RF soils in Indian tropical environments is available with benchmark status for many of these soils by the courtesy of the ICAR-NBSS&LUP, Nagpur, India (Lal et al. 1994; Murthy et al. 1982; Bhattacharyya et al. 2009). Over the past two decades however, the focus of

Table 3.1 Distribution of RF soils in agro-ecological regions, major physiography, parent material and Indian states and their taxonomic soil orders

Agro-ecological region (AER)[a] no.	Kind of AER[a]	Mean annual rainfall (MAR) mm[a]	Major physiography and states[a]	Parent material[a]	States[b]	Soil order, in order of abundance[b]
AER 7	Semi-arid ecosystem (*Hot semi-arid ecoregion in Eastern Ghats and Deccan plateau*)	600–1000	Andhra Pradesh and Eastern Ghats	Granite gneiss, Calcic gneiss	Andhra Pradesh	Alfisols, Inceptisols, Entisols, Mollisols
AER 8	Semi-arid ecosystem (*Hot semi-arid ecoregion in Eastern Ghats and Deccan plateau*)	600–1100	Eastern Ghats, Tamil Nadu Upland, Karnataka Plateau	Granite gneiss, Calcic gneiss	Karnataka	Alfisols, Inceptisols
				Granite gneiss, Calcic gneiss	Tamil Nadu	Alfisols, Inceptisols
AER 12	Sub-humid ecosystem (*Hot sub-humid ecoregion in Eastern Plateau and Eastern Ghats*)	1000–1800	Eastern Plateau (Chota Nagpur), Eastern Ghats of Orissa, Bihar, West Bengal	Ferruginous Sand stone	Bihar	Alfisols, Inceptisols, Entisols
				Ferruginous Sand stone	Orissa	Alfisols, Entisols
				Ferruginous Sand stone	West Bengal	Alfisols,
				Basalt	Madhya Pradesh	Alfisols, Mollisols
AER 15	Hot-per-humid ecosystems (*Hot sub-humid moist to humid with inclusion of per-humid ecoregion*)	1400–2000	Assam, Tripura, West Bengal Plains	Ferruginous Sand stone	Assam	Alfisols, Ultisols, Entisols, Inceptisols
				Ferruginous Sand stone	Tripura	Ultisols, Inceptisols, Alfisols

(continued)

Table 3.1 (continued)

Agro-ecological region(AER)[a] no.	Kind of AER[a]	Mean annual rainfall (MAR) mm[a]	Major physiography and states[a]	Parent material[a]	States[b]	Soil order, in order of abundance[b]
AER 17	Humid-Per-humid ecosystems (*Warm per-humid ecoregion in north-eastern hills*)	2000–3000	North-eastern hills of Nagaland, Mizoram, Manipur, Meghalaya, Arunachal Pradesh, Tripura	Gneiss, alluvium	Sikkim	Inceptisols, Entisols
				Gneiss, alluvium	Manipur	Ultisols, Inceptisols, Alfisols
				Gneiss, alluvium	Nagaland	Inceptisols, Ultisols, Alfisols, Entisols
				Gneiss, alluvium	Mizoram	Ultisols, Inceptisols, Alfisols, Entisols
					Meghalaya	Ultisols, Inceptisols, Alfisols, Entisols
				Gneiss, alluvium	Arunachal Pradesh	Inceptisols, Ultisols, Entisols, Alfisols

(continued)

Table 3.1 (continued)

Agro-ecological region(AER)[a] no.	Kind of AER[a]	Mean annual rainfall (MAR) mm[a]	Major physiography and states[a]	Parent material[a]	States[b]	Soil order, in order of abundance[b]
AER 18	Coastal ecosystems (*Hot sub-humid to semi-arid ecosystem in eastern coastal plain*)	900–1600	Coastal Andhra Pradesh, Orissa, Pondicherry, Tamil Nadu, West Bengal	Sandstone	Puducherry	Alfisols
AER 19	Coastal ecosystem (*Hot humid-per-humid ecoregion in Western Ghats and coastal plain*)	>1800	Western Ghats, coastal lands of Gujarat, Karnataka, Kerala, Maharashtra, Tamil Nadu, Goa, Daman and Diu	Laterite	Gujarat	Alfisols
				Laterite	Kerala	Ultisols, Alfisols, Mollisols, Inceptisols
				Granite-gneiss	Karnataka[c,d]	Ultisols, Mollisols, Alfisols
				Granit-gneiss	Tamil Nadu[e]	Ultisols, Alfisols, Mollisols
				Granite-gneiss[f]	Goa	Ultisols, Alfisols
				Basalt	Maharashtra	Alfisols, Inceptisols, Mollisols
AER 20	Island ecosystem (*Hot humid to per-humid island ecoregion in islands of Andaman and Nicobar*)	3000	Andaman and Nicobar islands	Calcareous sandstone, limestone, Micaceous sandstone[g]	Andaman and Nicobar islands	Mollisols, Alfisols

[a]Adapted from Sehgal et al. (1998); [b]Bhattacharyya et al. (2009), [c]Shiva Prasad et al. (1998), [d]Kharche, (1996); [e]Natarajan et al. (1997); [f]Harindranath et al. (1999); [g]Das et al. (1996)

research has shifted from general pedology to mineralogical and micro-morphological and climate change research (Pal et al. 2014). Although the present treatise is based on research results obtained from selected benchmark Indian tropical RF soils, data from other tropical parts of the world are also included where relevant.

This treatise addresses the changes in soil properties as a result of climatic shift during the Holocene. Such changes in soil properties more importantly a general trend in their degradation are especially exacerbated with a rise in mean annual temperature (MAT) and decrease in mean annual rainfall (MAR). Indeed, climate change effects on soils in the semi-arid tropical (SAT) regions are observed in their chemical degradation as a result of increased formation of pedogenic $CaCO_3$ and concomitant increase in subsoil sodicity, and the pedogenetic processes related to degradation is detailed in Chap. 6. The regressive pedogenic processes observed in vast areas of SAT red ferruginous soils (Table 3.1) manifests as decreased crop productivity under rain fed agricultural systems (Pal et al. 2013). This general degradation of soil resource needs urgent attention for developing management strategies to rejuvenate the degraded soil resource base as the livelihoods of a large number of small and poor farmers depends on these soils (Pal et al. 2014). The main objective of this treatise is to integrate the available information on the pedology of RF soils so that the land resource managers can take this as guiding principles in their efforts to optimize the soil productivity for the food security in India and other tropical regions in the 21st century.

3.2   US Soil Taxonomic Classes and Factors in the Formation of Red Ferruginous (RF) Soils

The RF soils are developed on many rock systems, and they are mostly on granite, granite-gneiss, ferruginous sandstone and schist of Achaean Period, and in places Deccan basalt, shale, limestone and marine deposits (Sehgal et al. 1998; Bhattacharyya et al. 2009) (Table 3.1). As they have thermic, isothermic, hyperthermic and isohyperthermic soil temperature regimes, their soil moisture regimes are thus ustic, udic and aquic. It is interesting to note that although the soils belong to Entisols, Inceptisols, Alfisols, Ultisols and Mollisols orders of the US Soil Taxonomy (Bhattacharyya et al. 2009), their soil orders are not similar under all soil moisture and temperature regimes. Under humid tropical (HT) climates, they are Inceptisols, Alfisols, Mollisols, and Ultisols in Kerala; Inceptisols, Alfisols, and Ultisols in Goa; Alfisols, Mollisols, and Ultisols in Karnataka; Alfisols, Mollisols, and Ultisols in Tamil Nadu; Inceptisols, Alfisols and Mollisols in Maharashtra; Alfisols in Gujarat; Entisols and Inceptisols in Sikkim; Entisols, Inceptisols, Alfisols and Ultisols in Arunachal Pradesh; Entisols, Inceptisols, Alfisols and Ultisols in Assam; Inceptisols, Alfisols and Ultisols in Manipur; Entisols, Inceptisols, Alfisols and Ultisols in Mizoram; Entisols, Inceptisols, Alfisols and

Ultisols in Meghalaya; Entisols, Inceptisols, Alfisols and Ultisols in Nagaland; Inceptisols, Alfisols and Ultisols in Tripura; and Alfisols and Mollisols in Andaman and Nicobar Islands (Pal et al. 2014; Table 3.1).

Under relatively dry bio-climates like in arid, semi-arid and sub-humid, RF soils also belong to different soil orders. They are Entisols, Inceptisols, Alfisols and Mollisols in Andhra Pradesh; Inceptisols and Alfisols in Karnataka; Inceptisols and Alfisols in Tamil Nadu; Entisols, Inceptisols and Alfisols in Bihar; Entisols and Alfisols in Orissa; Alfisols and Mollisols in Madhya Pradesh; Alfisols in West Bengal; and Alfisols in Puducherry (Pal et al. 2014; Table 3.1). Such taxonomic diversity in RF soils of India appears while intriguing, is however, a unique natural endowment, and thus suggests that soil diversity in the geographic tropics of India is as large as that observed in the temperate regions (Eswaran et al. 1992a; Sanchez 1976; Sanchez and Logan 1992). Consequently, generalizations on the management of RF soils would be rather imprudent as they are unlikely to have wider applicability because of a large diversity in these soils as a result of their genesis leading to a wide range in physical, chemical and biological properties (Pal et al. 2014). Therefore, uniqueness of the diversity of RF soils caused by parent material composition, the presence of soil modifiers, climate and vegetation, demands a careful attention of the pedologists in better understanding of the subtle pedogenetic processes, which are dealt in the following sections.

3.3   Clay Illuviation, Clay Enriched B Horizons and Taxonomic Rationale of RF Soils of Humid Tropical (HT) Climate

Although they are mild to strongly acidic in nature, the Ultisols, Alfisols and Mollisols have clay enriched B horizons. Texture in these soils varies from fine loamy to clayey (Bhattacharyya et al. 2009). It is interesting to note that despite the acidity, Alfisols and Mollisols maintain the required base saturation, because of the enrichment by basic cations of the parent material derived from calc-gneiss rock and limestone (Pal et al. 2014). Identification of the argillic horizon in Indian Ultisols has not been always straightforward in the field (Bhattacharyya et al. 1994; Sen et al. 1994) and elsewhere (Beinroth 1982; Eswaran 1972; Rebertus and Buol 1985). On many occasions this has led to their inappropriate place in the US Soil Taxonomy by undermining possibly the major pedogenetic process on a stable geomorphic surface.

To circumvent this problem the US Soil Taxonomy introduced the 'Kandic' concept (Soil Survey Staff 1990). This concept aptly provides a scope to eliminate the requirement for argillic horizon in Ultisols, which are generally low activity clay (LAC) soils. Many Ultisols of Kerala (Eswaran et al. 1992b; Krishnan et al. 1996), Tamil Nadu (Natarajan et al. 1997), Karnataka (Eswaran et al. 1992b; Kharche 1996; Shiva Prasad et al. 1998) in the southern peninsular area and of Arunachal

Pradesh, Assam, Meghalaya, Nagaland, Tripura (Sen et al. 1997a) and Manipur (Sen et al. 1994) in the northeastern hills (NEH) have sub-surface horizons that meet both the textural and depth requirement of the Kandic horizon. Surface soils are strong to moderately acidic but the sub-surface soils are very strongly acidic resulting in poor base status. Surface soils have >0.9% organic carbon (OC), which decreases with depth. The CEC (by 1 N NH_4OAc, pH 7) and ECEC (sum of bases extracted with 1 N NH_4OAc, pH 7 plus 1 N KCL extractable Al_3^+) of clay in the subsurface horizons are less than 16 cmol (p+) kg^{-1}even though the soils are clayey and rich in OC. Therefore, such soils qualify for LAC soils and lend support to the existence of Kandic horizon. It is intriguing to note that many non-LAC soils of NEH have no lithological discontinuity but have requisite clay enrichment in the argillic horizon without identifiable clay skins in the field (Sen et al. 1997b). Similar observations have also been reported earlier for the semi-arid soils of the Indo-Gangetic Alluvial Plains (IGP). But a detailed micro-morphological study (Pal et al. 1994) confirmed the occurrence of clay illuviation process in these soils. Due to the lack of thin-section studies' support to confirm the clay illuviation process, such acid soils of NEH with clay enriched B horizons have hitherto been placed under Dystrochrepts (Sen et al. 1997b) by undermining a major pedogenic process operating in these soils for a long geological period on a stable landscape. This warrants a detailed micro-morphological study of acidic pH soils of India to investigate the precise cause–effect relation of the presence/absence of clay skins (that are generally identified by 10× hand lens in the field).

Limited data on the clay pedofeatures in such soils (particularly Ultisols and soils with Ultic characters) in India (Kooistra 1982) and elsewhere (Eswaran and Sys 1979) indicate the presence of pure void argillans (Fig. 3.1a), which however are not considered to be the result of current pedogenic processes (Eswaran and Sys 1979; Kooistra 1982). The dispersion of clay particles is possible under slightly acidic to moderately alkaline pH conditions at a very low electrolyte concentration that ensures a pH higher than that required for the zero point of charge for complete dispersion of clay (Eswaran and Sys 1979). Therefore, the movement of defloc-culated clays (and its subsequent accumulation in the Bt horizons) occurs at the initial stage of soil formation when the pH is moderately alkaline and remains above the point of zero charge. Such chemical environment would result in deflocculation, disengaging face-to-face association of clay particles (Van Olphen 1966); and consequently cause impairment of parallel orientation of the clay pla-telets. In this colloidal state, the illuviation of the fine clay particles would result in textural pedofeatures of the 'impure type' (Fig. 3.1b) as observed in soils of the north-western (NW) parts of the IGP during the Holocene (Pal et al. 1994).

The presence of pure void argillans reported in some RF soils suggests that with the advancement in weathering and leaching of bases, and concomitant lowering of pH during the initial stages of pedogenesis in HT climate, clay platelets could remain in face to face association or parallel, or oriented aggregation when the flocculation of the clay suspension was not induced by the presence of salts (Van Olphen 1966), particularly the carbonates and bicarbonates of sodium (Pal et al.

(a)   (b)

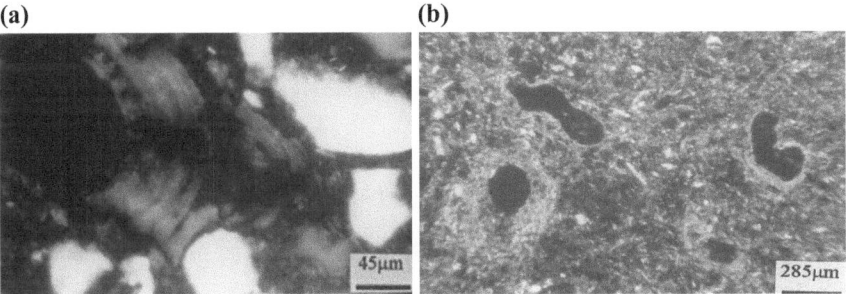

Fig. 3.1 Thin-section photographs showing typical micromorphological features of illuvial clay pedofeatures showing micro laminations and strong preferred orientation of ferruginous soils of southern India (**a**) (Adapted from Venugopal et al. 1991), and impure clay pedofeatures in soils of the NW part of the IGP (**b**) (Adapted from Pal et al. 2003)

1994). Therefore, the clay illuviation in such highly weathered acidic soils of HT climate is not a current pedogenetic process (Eswaran and Sys 1979; Kooistra 1982). This is further supported by the presence of granular microstructure, organic matter rich flocculated clay granules, and thin-to-thick ferruginous clay coatings in these soils as observed by Kooistra (1982), Bockheim and Hartemink (2013).

3.4   RF Soils (Alfisols) in Semi-arid Tropics (SAT): Record of Climate Change and Polypedogenesis in the Holocene Period

Laterite mounds and laterite plateau remnants are scattered over the landscape in parts of southern and eastern India. In central and western India thin-to-thick (0.25–3.00 m) laterite cappings occur on various rock types ranging in ages from Archean to Gondwanas. Large numbers of massive granitic tors in gneissic terrain bear the evidence of exhumation during the dry period following prolonged deep weathering in the HT climate that prevailed from the Upper Cretaceous until Plio-Pleistocene (Pal et al. 2014). The Plio-Pleistocene was a transition period when the climate became drier with rising of the Western Ghats (Brunner 1970). As a result, the upper layers of RF soils (mainly Paleustalfs and Rhodustalfs) formed in the preceding HT climate, were truncated by multiple arid erosional cycles (Pal et al. 2014). Due to truncation of the upper layers, the coarse and fine clay contents presently show an upward increase in the solum, a sharp decline in the Ap horizon (Fig. 3.2a) and argillans immediately beneath the Ap horizon. Such modification in the geomorphic surface is also evident from the presence of broken argillans (papules) (Pal et al. 2014) in the soil solum, which possibly was caused by low-energy short distance transport in the upland undulating areas (Srivastava et al. 2010).

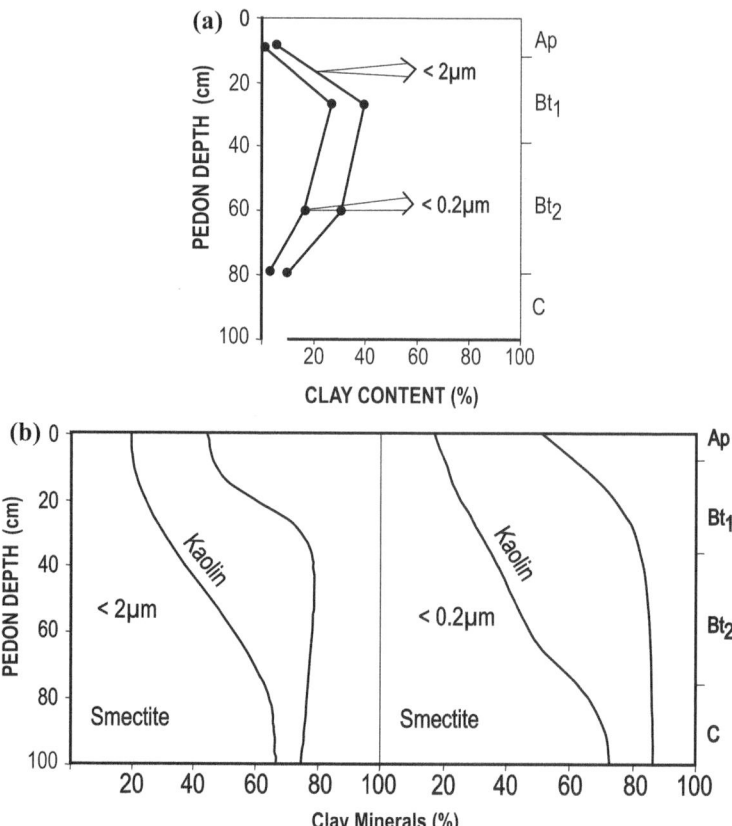

Fig. 3.2 Depth distribution of clay fraction (**a**) and clay size smectite and kaolin (**b**) with pedon depth of a representative SAT Alfisol (Adapted from Pal et al. 2000a; Pal and Deshpande 1987)

The texture of most of the SAT Alfisols ranges from fine loamy to clayey (Bhattacharyya et al. 2009), and such soils are in general calcareous and almost neutral to mildly alkaline in nature. These Alfisols are spatially associated with Vertisols (Typic/Aridic/Sodic Haplusterts, Pal et al. 2012) and contain both dioctahedral smectite and kaolin (0.7 nm mineral) (Bhattacharyya et al. 1993) in their clay fractions alongside mica (Pal and Deshpande 1987; Pal et al. 1989; Chandran et al. 2000). Such selected soils however, are dominated by kaolin with small amount of smectite (Pal and Deshpande 1987). In the Alfisol pedon, kaolin decreases and smectite increases with depth, and the sum of these two minerals are fairly constant from the B to the C (saprolites) horizon (Fig. 3.2b), which suggests the transformation of smectite to kaolinite through an intermediate phase of smectite–kaolinite (Sm–Kl) interstratifications identifiable by broad basal reflections of kaolinite at 0.74–0.75 and 0.35 nm (Bhattacharyya et al. 1993) (Fig. 3.3). However, in few bench mark SAT Alfisols of the Mysore plateau in Karnataka

(Murthy et al. 1982), 0.7 nm mineral is a fairly well crystalline kaolinite mineral (XRD diagram shown in inset, Fig. 3.4b) as the weathered product of biotite (Fig. 3.5a), and is observed as hexagonal crystals under SEM as pseudomorphs after plagioclase (Fig. 3.5b). The calculated unit cell cation composition of such kaolinites showed a substitution of 0.11–0.82 atoms of Fe (III) for Al in every four octahedral sites (Rengasamy et al. 1975). Schwertmann and Herbillon (1992) suggested that such substitutions inhibit crystal development but make the kaolinite more active with higher surface area.

For Alfisols of Jharkhand, Ray et al. (2001) observed that kaolinites have CEC more than 30 cmol (p+) kg^{-1}. In such SAT Alfisols, the dominance of smectite alongside small amount of kaolinite throughout the solum is also quite common (Chandran et al. 2000). These observations indicate that SAT Alfisols have both smectite and 0.7 nm clay minerals that are basically related to quite contrasting chemical environments, indicating the influence of humid and SAT climates at different geological times in the past. Present SAT climatic conditions are not considered severe enough for transformation and weathering to kaolinite and also for the formation of huge amount of smectite in ferruginous and spatially associated Vertisols (Pal et al. 2012). Moreover, neutral to alkaline reactions of the present day

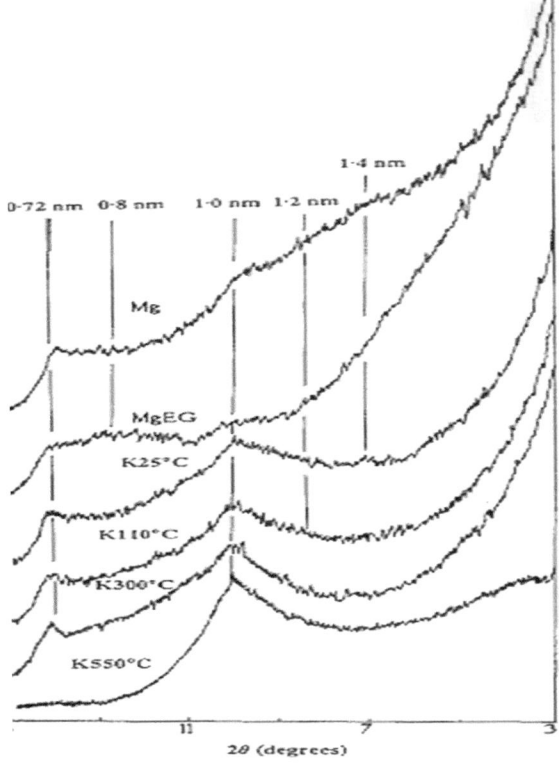

Fig. 3.3 X-ray diffractogram of the 0.72 nm and HIS minerals in the fine clays of the representative Alfisols developed on zeolitic Deccan basalts of HT climate (Adapted from Bhattacharyya et al. 1993)

(a)

(b)

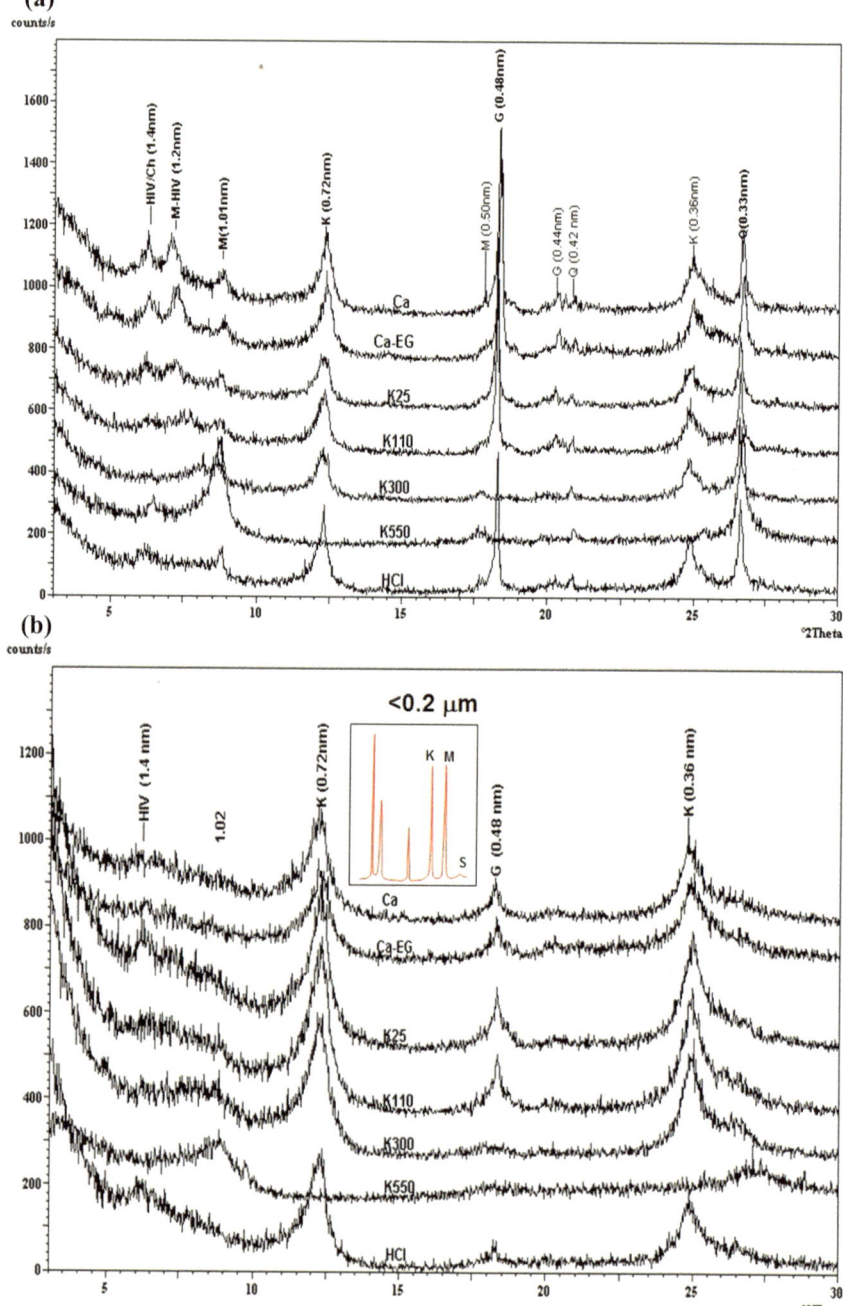

◀ **Fig. 3.4** Representative XRD diagrams of total clay (**a**) and fine clay (**b**) fractions of Ultisols of HT climate: Ca, Ca-saturated; Ca-EG, calcium saturated and ethylene glycolated; K25, K110, K300, K550, K-saturated and heated to 25, 110, 300, 550 °C, respectively; HCl, treated with 6 N HCl for 30 min at 90 °C. HIV, hydroxy-interlayered vermiculite; M-HIV, mica-HIV minerals; *M* mica; *K* kaolin; *Ch* chlorite; *Q* quartz, *G* gibbsite. Well crystalline fine clay kaolinite mineral is shown in the *inset* (**b**) for comparison to kaolin (Adapted from Chandran et al. 2005a; Pal et al. 1989)

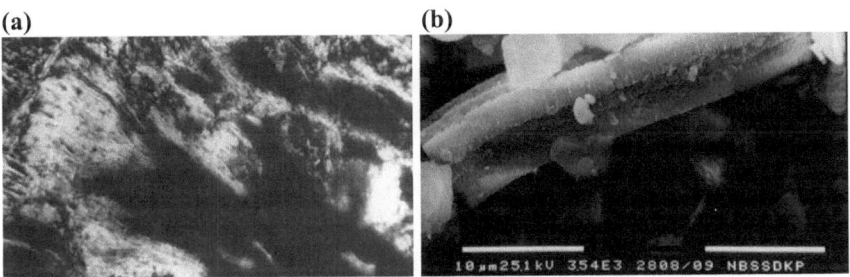

(**a**)   (**b**)

Fig. 3.5 Partially altered biotite associated with kaolinite pseudomorphs after biotite (**a**), and kaolinite pseudomorphs after feldspar of SAT Alfisols of Mysore plateau of Karnataka (**b**) (Adapted from Pal et al. 1989; Pal and Sarma 2002)

soils do not favour the transformation of smectite to kaolinite. Therefore, kaolinite was formed in the earlier pre-Pliocene geological period with higher rainfall and greater fluctuations in temperature (Pal et al. 2014).

It is interesting to note that the dioctahedral smectite was the first weathering product of peninsular gneiss during the HT climate of pre-Pliocene period, but its stability was ephemeral in this climate as evident from its transformation to kaolinite. Both Sm–Kl and smectite could be preserved in such soils because of termination of HT climate during the Plio-Pleistocene transition (Pal et al. 1989). Such SAT Alfisols are spatially associated with Vertisols in the lower topographic positions, and these Vertisols are developed during the dry climate of the Holocene period in the deposited smectitic parent material of the previous much wetter climate (Pal et al. 2009). The preservation of the crystallinity of SAT Vertisols' smectite in depressions and the lack of transformation of primary minerals validate the hypothesis of positive entropy change during the formation of Vertisols (Srivastava et al. 2002). Therefore, the formation of such Vertisols acts as a tool to infer climate change in geographical areas covered by the RF soils in the Peninsular India (Pal et al. 2012).

In view of the formation of 1.7 and 0.7 nm clay minerals in the previous HT climate, RF soils of the Peninsular India overlying the saprolites of metamorphic rocks, dominated either by kaolinite or dioctahedral smectite, are relict paleosols (Pal et al. 1989). These relict soils have been affected by the climatic change from humid to semi-arid tropical (SAT) climatic conditions during the Plio-Pleistocene transition period as evident from the formation of trioctahedral smectite during the SAT climate from the sand and silt size biotite (Fig. 3.6a), which survived earlier

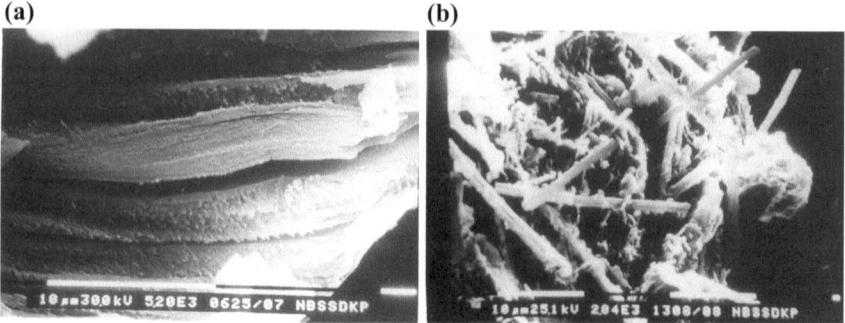

Fig. 3.6 Biotite and biotite interstratified with vermiculite survived further weathering of HT climate (**a**) and the cluster of calcite needles (**b**) in the present SAT Alfisols of Karnataka (Adapted from Pal et al. 2000a)

weathering during the HT climate (Pal et al. 2014). In addition, few such soils show the presence of pedogenic calcium carbonates (PC) as spongy nodules and cluster of carnonets needles below 40 cm depth (Fig. 3.6b), indicating their formation in the prevailing SAT environments.

The formation of PC causes the concomitant development of sodicity (Pal et al. 2000b). At present, the rate of formation of PC and the development of sodicity (ESP) in the SAT Alfisols may not be alarming. However, the continuation of the formation of PC and ESP as natural chemical degradation processes would finally impair the drainage of these soils (Pal et al. 2000b; Srivastava et al. 2001). During the SAT climate for millions of years many such soils are at present mildly acidic and calcareous (Bhattacharyya et al. 2009), and their clay enriched B horizons often indicate the presence of impure clay pedofeatures (Venugopal 1997) along with pure void argillans (Kooistra 1982; Venugopal et al. 1991; Venugopal 1997). In the SAT climate, the precipitation of $CaCO_3$ facilitates the deflocculation of clay particles and their subsequent translocation and accumulation in the Bt horizons (Pal et al. 2003), but in the presence of various types of other illuvial pedofeatures (Venugopal 1997), it would not be prudent to suggest a precise pedogenic process for the formation of poorly oriented (impure) clay pedofeatures. Fresh research initiative in this area of pedology is thus warranted.

3.5   Acidity and Its Nature and Charge Characteristics in RF Soils of HT Climate: The Present Status

High rate of mineral weathering in HT climate of India and the excessive leaching of soluble salts cause acidity in the surface and sub-surface of soils, and thus soil acidity poses a challenge for crop production as the soils are slight to strongly acidic in pH. Acidity develops on soil colloids mainly through isomorphous substitution of H^+ or Al_{3+} in silicate minerals, which form exchange sites throughout the pH ranges (permanent charge acidity i.e. the exchangeable acidity). Soil acidity is also developed due to the formation of polymers of Fe and Al with soil organic matter when the exchange sites mainly depend mainly on soil pH i.e., the pH-dependent acidity. Research results of such soils of Kerala (Chandran et al. 2005a; Krishnan et al. 1996), Goa (Chandran et al. 2004), Karnataka (Shiva Prasad et al. 1998), Tamil Nadu (Natarajan et al. 1997) and NEH (Bhattacharyya et al. 1994, 2000, 2004; Nayak et al. 1996a; Sen et al. 1994, 1997a, b, c) indicate that KCl extractable acidity is low but total acidity by $BaCl_2$-TEA is remarkably high (Table 3.2). It is interesting to note that exchange acidity contributes up to 22%, and the pH-dependent acidity contributes about 98% of the total acidity. Such high amount of pH-dependent acidity is due to the combined effects of organic matter, free sesquioxides and terminal hydroxyl groups associated with Si and/or Al, and/or the ruptured Si-O-Al bonds (associated mostly with low activity clays) (Nayak et al. 1996a; Sen et al. 1997a).

   With decrease in soil pH, the exchangeable $Al_3{}^+$ and Al-saturation increase, and it is almost nil at about pH around 6.0. Values of the KCl exchangeable acidity in some Ultisols of Kerala (Chandran et al. 2005a), Meghalaya (Bhattacharyya et al. 1994, 2000) and Alfisols of Goa (Chandran et al. 2004) is often less than 1 cmol $(p+)$ kg^{-1}, and the KCl pH values of these soils in the lower horizons are close to or greater than pH values in water (Table 3.2), suggesting the presence of gibbsite and/or amorphous materials (Smith 1986). Beinroth (1982) made similar observations in highly weathered Oxisols of Puerto Rico with gibbsite, and he believed that these Ultisols are in the advanced stage of tropical weathering because such low values of exchange acidity are often thought to be associated with the presence of gibbsite minerals. However, Bhattacharyya et al. (2000) and Chandran et al. (2005a) demonstrated that the formation of gibbsite is improbable during the highly acidic weathering conditions of HT climate; and therefore it cannot be taken as an index of advanced stage of tropical weathering. It is interesting to note that in non-gibbsitic Ultisols and Inceptisols of NEH, KCL exchangeable acidity is 3–4 times higher values than unity (Bhattacharyya et al. 1994; Nayak et al. 1996a; Sen et al. 1997a).

   Chandran et al. (2004, 2005b) indicated that the low amount of KCl extractable acidity (due to H^+ and $Al_3{}^+$ ions) is due to release of considerable amount of $Al_3{}^+$ during humid tropical weathering, which is trapped in clay minerals other than kaolinite. Such trapped $Al_3{}^+$ is not easily extractable by 1 N unbuffered KCl solution. The LAC soils have kaolinitic mineralogy class according to the CEC of soils (<16 cmol $(p+)$ kg^{-1}) and ECEC of clays (<12 cmol $(p+)$ kg^{-1}) (Bhattacharyya et al. 1994; Eswaran et al. 1992b; Kharche 1996; Krishnan et al. 1996; Natarajan et al. 1997; Nayak et al. 1996; Sen et al. 1994, 1997a, b, c; Shiva Prasad et al. 1998).

Table 3.2 Physical and chemical properties of selected RF soils of HT climate

Horizon	Depth (cm)	pH H₂O	pH KCl	ΔpH	Clay (%)	Organic carbon %	Exchange acidity KCl (N) H⁺ cmol(P⁺)/keg	Al³⁺	Extractable acidity, BaCl₂-TEA	CEC, soil (NH₄OAC.7)	ECEC, soil (NH₄OAC.7)	CEC, clay	ECEC, clay	Base saturation (%)	Lime requirement (t/ha)
Ustic Kandihumults: Kerala[a]															
Ap	0–13	4.8	4.3	−0.5	21.1	2.35	0.33	0.60	11.2	4.5	1.6	21.3	7.6	22	–
Bt1	13–32	4.4	4.3	−0.1	31.3	1.86	0.50	0.48	10.4	3.5	1.1	11.2	3.5	17	–
Bt2	32–56	4.5	4.3	−0.2	29.0	1.50	0.43	0.32	10.0	3.7	0.8	12.8	2.7	14	–
Bt3	56–83	4.5	4.6	+0.1	26.0	0.90	0.23	0.12	9.0	4.1	0.7	16.0	2.7	13	–
Bt4	83–112	4.4	4.6	+0.2	28.5	1.11	0.20	0.10	6.6	4.0	0.6	13.7	2.1	13	–
Bt5	112–150	4.7	4.7	Nil	24.0	1.22	0.30	0.10	7.0	4.0	0.7	16.7	3.0	17	–
Kanhaplohumults: Arunachal Pradesh[b]															
A1	0–13	4.6	3.8	−0.8	53.0	2.70	2.0	0.7	23.3	13.2	4.8	23.0	9.2	16	3.6
Bt1	13–36	4.8	3.8	−0.7	66.0	1.96	2.2	1.0	18.5	12.3	4.1	18.6	6.2	18	4.6
Bt21	36–63	5.0	4.5	−0.8	68.0	0.87	2.4	1.0	14.2	10.5	4.0	15.4	6.0	16	5.1
Bt22	63–100	5.4	4.2	−0.8	65.0	0.64	2.6	1.1	14.3	9.7	4.4	15.0	6.8	17	5.4
B3	100–125	5.2	4.2	−1.0	47.5	0.29	2.7	1.1	10.2	7.1	4.4	15.0	6.8	18	5.4
Kanhaplohumults: Assam[b]															
A1	0–14	5.0	3.8	−1.2	48.5	2.20	1.4	0.4	16.1	11.0	5.6	22.4	11.5	34	2.1
Bt11	14–33	4.8	3.7	−1.1	56.5	1.60	2.2	0.9	16.0	10.5	5.6	18.6	10.0	24	4.3
Bt12	33–85	5.1	3.8	−1.3	62.0	1.00	2.0	0.9	13.0	9.0	4.4	14.3	7.1	17	4.5
Bt21	85–120	5.2	4.0	−1.2	65.5	0.70	1.5	0.6	12.2	7.7	4.0	11.7	6.0	21	3.2
BC	120–180	5.4	4.0	−1.4	63.5	0.50	1.3	0.5	12.0	6.5	3.1	10.2	5.0	19	2.6

(continued)

Table 3.2 (continued)

Typic Dystrochrepts: Assam[b]

A1	0–13	5.0	3.9	−1.1	29.5	1.50	1.4	0.3	10.7	10.0	5.4	32.5	18.2	37	1.3
A2	13–31	4.8	3.9	−0.9	34.5	0.80	2.3	0.9	9.3	7.5	5.5	21.7	16.0	30	4.3
B21	31–70	4.9	3.9	−1.0	34.0	0.66	2.1	0.7	10.6	7.0	4.8	20.3	14.0	22	3.6
B22	70–122	4.7	4.1	−0.6	44.5	0.57	2.1	0.7	10.7	8.1	4.6	18.2	10.0	18	3.6
B3	122–175	5.0	4.1	−0.9	29.0	0.32	1.5	0.4	11.6	5.3	3.3	18.2	12.2	29	2.2

Typic Dystrochrepts: Manipur[b]

A1	0–10	4.0	3.6	−0.4	35.0	1.60	2.4	1.1	10.5	9.2	7.8	26.3	22.3	51	5.3
A2	10–28	4.0	3.6	−0.4	40.0	1.20	2.9	1.4	8.7	8.1	6.3	20.2	15.7	28	6.7
B21	28–60	4.2	3.8	−0.4	40.5	0.70	3.3	1.5	8.8	8.0	5.8	19.5	14.3	26	7.6
B22	60–90	4.2	3.8	−0.4	39.0	0.60	3.1	1.5	8.6	7.6	5.5	19.5	14.0	30	7.4
BC	90–125	4.2	3.9	−0.3	28.0	0.30	2.7	1.3	6.5	7.8	5.0	27.8	17.5	13	6.4

Typic Kandihumults:Meghalaya[b]

A1	0–16	4.5	4.2	−0.3	21.6	3.6	1.14	0.36	32.0	7.3	2.5	33.8	16.6	19	1.8
B21	16–31	4.8	4.4	−0.4	30.4	2.5	0.64	0.16	27.0	7.3	3.0	24.0	10.0	20	0.8
B22	31–62	5.0	4.8	−0.2	31.1	2.0	0.30	Nil	22.0	4.1	0.8	13.2	2.6	19	Nil
B23	62–95	5.1	6.4	+0.9	26.5	0.6	0.10	Nil	15.8	4.1	1.2	15.5	4.5	29	Nil

Typic Hapludults:Mizoram[b]

A1	0–27	4.7	3.8	−0.9	37.0	1.00	3.27	1.53	14.3	10.6	6.8	28.6	18.3	19	7.6
Bt1	27–50	4.7	3.8	−0.9	45.0	0.80	3.43	1.57	12.7	10.0	6.2	22.7	14.5	13	7.7
Bt2	50–110	4.9	3.8	−1.1	50.0	0.60	3.53	1.67	14.3	10.3	6.6	20.6	13.2	14	8.2
BC	110–180	5.0	3.9	−1.1	36.5	0.40	2.33	1.07	12.0	7.3	5.3	20.0	14.3	19	5.3

(continued)

Table 3.2 (continued)

Typic Hapludults: Nagaland[b]

A1	0–15	5.5	4.1	−1.4	33.0	1.40	0.54	0.06	10.3	11.4	8.0	34.5	24.4	65	0.3
Bt1	15–30	5.1	3.6	−1.5	44.0	0.70	1.74	0.76	12.0	10.0	7.0	22.7	16.1	47	3.8
Bt21	30–46	5.1	3.6	−1.5	55.5	0.60	3.00	1.40	12.4	12.4	9.3	22.3	16.7	39	6.9
Bt22	46–71	5.2	3.5	−1.7	68.0	0.50	4.33	2.07	15.0	14.5	11.6	21.3	17.0	35	10.2
Bt31	71–100	5.2	3.5	−1.7	67.5	0.40	4.84	2.36	17.0	15.0	12.1	22.0	20.0	38	11.7
Bt32	100–125	5.3	3.5	−1.8	64.5	0.30	4.70	2.30	17.1	13.3	12.0	20.6	18.6	46	11.4

Typic Paleudults: Tripura[b]

A1	0–11	4.7	3.9	−0.7	18.5	0.70	1.5	0.67	7.3	5.8	4.1	31.3	22.2	33	3.3
Bt1	11–38	4.5	3.8	−0.7	29.5	0.60	2.3	1.13	7.4	6.6	4.9	22.4	16.3	21	5.6
Bt2	38–83	4.6	3.9	−0.7	35.5	0.40	2.3	1.10	8.8	7.0	4.9	19.7	13.9	22	5.4
Bt3	83–172	4.8	3.9	−0.9	36.0	0.30	2.2	1.03	10.9	7.4	4.5	20.5	12.5	17	5.1

Kandic Paleustalfs: Goa[c,d]

A	0–8	5.4	4.6	−0.8	46.2	5.04	0.17	0.05	19.0	19.0	7.6	41.1	16.4	40	–
A2	8–19	5.3	4.4	−0.9	62.2	2.52	0.19	0.37	19.6	15.7	6.8	25.0	11.0	41	–
Bt1	19–51	5.2	4.4	−0.8	57.2	2.01	0.21	0.41	18.0	16.2	6.6	28.3	11.5	38	–
Bt2	51–89	5.2	4.4	−0.8	65.2	1.47	0.22	0.38	16.5	14.3	6.9	22.0	10.6	45	–
Cr	89+	–													

Pachic Argiustoll: Kerala[e]

A	0–19	5.1	–	–	22.7	1.80	–	–	–	15.4	–	67.8	–	64	–
Bt1	19–53	6.4	–	–	22.6	1.28	–	–	–	13.7	–	60.6	–	61	–
Bt2	53–81	6.4	–	–	23.0	0.38	–	–	–	10.8	–	47.0	–	74	–
Bt3	81–115	6.5	–	–	20.8	0.40	–	–	–	9.7	–	46.6	–	73	–
Bt4	115–155	6.4	–	–	22.7	0.34	–	–	8.82	8.6	–	38.0	–	76	–

[a]Adapted from Chandran et al. (2005a); [b]Adapted from Sen et al. (1997c); [c]Adapted from Chandran et al. (2004); [d]Adapted from Harindranath et al. (1999); [e]Adapted from Krishnan et al. (1996), – = data not available

3.6   Clay Mineralogy of RF Soils of the HT Climate: Recent Advances

For the last few decades many researchers believed that tropical soils as those of the hot and humid tropics, exemplified by deep red and highly weathered soils (Eswaran et al. 1992a) do not efficiently support plant growth and have low productivity (Aleva 1994) because of their kaolinitic/gibbsitic mineralogy (Schwertmann and Herbillon 1992) associated with Fe and Al toxicity (Sehgal 1998). But, recent research using high resolution X-ray diffraction (XRD) and scanning electron microscopy (SEM) on soil particle size fractions of the acidic bench mark RF soils (Ultisols, Alfisols and Inceptisols) of the HT climate of India in the states of Maharashtra (Bhattacharyya et al. 1993), Madhya Pradesh (Bhattacharyya et al. 2006), Karnataka (Kharche 1996), Kerala (including soils very close to Angadipuram-the type locality of laterite, a name first coined by Francis Buchanan in 1800, Chandran et al. 2005a), Goa (Chandran et al. 2004), Jharkhand (Ray et al. 2001), Meghalaya (Bhattacharyya et al. 2000), Tripura (Bhattacharyya et al. 2013), Manipur (Chandran et al. 2006), Assam (Pal et al. 1987) and Andaman and Nicobar islands (Chandran et al. 2005b) indicates the ubiquitous presence of kaolin (Kl-HIV/HIS) (a 0.7 nm mineral interstratified with hydroxy-interlayered vermiculite, HIV or smectite, HIS) as dominant clay mineral with occasional presence of gibbsite (Figs. 3.3 and 3.4a, b). The dominant presence of kaolin is, however, in contrast to the general perception that these soils are dominated by kaolinite and/or gibbsite. Interestingly, these soils also contain considerable amounts of weatherable minerals (>10%) like mica, mica-hydroxy-interlayered smectite (M-HIS), and hydroxy-interlayered vermiculite/smectite (HIV/HIS) (Chandran et al. 2004, 2005a). With this typical clay mineralogical make up any proposal on the presence of Oxisols order in the HT climate of the Indian subcontinent stands for careful re-examination.

3.6.1   Genesis of Gibbsite at the Expense of Kaolinite: Dissolution of a Myth

In view of the much accepted transformation pathway for the formation of gibbsite at the expense of kaolinite in text books on soil science, research endeavour at ICAR-NBSS&LUP during the last decade indicates that the XRD of Ca/Mg-saturated silt, total clay and fine clay fractions of Ultisols, Alfisols and Inceptisols of Indian HT climate on glycolation shows a characteristic 0.7 nm peak of kaolin, which has a broad base tailing towards the low angle side with branching at the tips (Figs. 3.3 and 3.4a, b). This 0.7 nm peak is not a peak of discrete kaolinite because on K-saturation and heating at 550 °C, the 0.7 nm peaks disappear reinforcing the 1.0 nm region of mica. Such reinforcement of 1.0 nm peak does not happen with pure kaolinite. The presence of mica, M-HIV (occasional),

HIS and HIV in the silt and clay fractions indicates that mica has been transformed into vermiculite first, then to HIV and finally to kaolin in soils developed from the metamorphic rocks (Bhattacharyya et al. 2000, 2013; Chandran et al. 2004, 2005a, b, 2006; Pal et al. 1987; Ray et al. 2001).

On basic rock formation genesis of kaolin from 2:1 clay minerals also followed the same pathway as discussed above. In RF soils developed on the Deccan basalt, smectite is first transformed to HIS and then to kaolin (Kl-HIS) (Bhattacharyya et al. 1993, 1999, 2005) (Fig. 3.3). The Al_{3+} released during tropical weathering is adsorbed in the interlayers of smectites to form HIS and further to kaolin (Kl-HIS). In an acid weathering environment induced by the HT climate with abundant Al_{3+}, hydroxy-Al-interlayering in the expansible clay minerals is the primary reaction towards the interstratifications of 2:1 and 1:1 layers (Pal et al. 1989; Bhattacharyya et al. 1997, 1997). Such interstratified minerals are common in RF soils of the present and past HT climate (Pal et al. 1989; Bhattacharyya et al. 1997). In Ultisols and Alfisols of HT climate of India, gibbsite and HIV/HIS occur simultaneously in contrast to an age long concept of 'antigibbsite effect' (Jackson 1964), which states that so long as the formation of HIV/HIS continues, formation of gibbsite is improbable. It is important to mention here that the presence of gibbsite is not a common occurrence in soils of the HT climate because it was only detected in Ultisols of Kerala (Chandran et al. 2005a), Alfisols of Goa (Chandran et al. 2004), Alfisols and Ultisols of Karnataka (Kharche 1996) and Ultisols of Meghalaya (Bhattacharyya et al. 2000). Such conspicuous absence of gibbsite in some highly weathered soils of the HT climate needs renewed research efforts.

Well-developed gibbsite as hexagonal crystals are observed under SEM, and they are pseudomorphs after feldspars (Fig. 3.7a, b) in Ultisols of Kerala, and sillimanite grains in Ultisols of Meghalaya (Fig. 3.7c). Interestingly, gibbsite is also a common occurrence in alkaline environment of natural bauxite (Balasubramaniam and Sabale 1984). Tait et al. (1983) reported the formation of gibbsite in an alkaline environment. It is an intriguing situation because the formation of kaolin is favoured in acidic environment while it is alkaline for gibbsite. Therefore, the formation of gibbsite and HIV/HIS occurs in two different chemical environments. The resultant clay mineral in an acidic environment after the prolonged tropical weathering is kaolin (Kl-HIV/HIS) and further desilication of this mineral is ceased at present because the pH is much below the threshold of $\sim$9 (Millot 1970). Incomplete desilication is evident from the higher ratios of $SiO_2:R_2O_3$ (1.4–5.0) and $SiO_2:A_2O_3$ (1.8–6.0) in Ultisols (Chandran et al. 2005a; Varghese and Byju 1993) and Alfisols (Chandran et al. 2004). Ultisols and acidic Alfisols are still siliceous, which suggests that the desilication process is not a part of the present pedogenic process in these soils of the HT climate of India (Pal et al. 2014).

Realizing the basic pedogenetic process of the siliceous but highly weathered acidic Alfisols and Ultisols of the Indian HT climate, it became necessary to follow the genesis of the reported occurrence of Oxisols of other tropical areas of the world (Chandran et al. 2005a; Pal et al. 2014). Surprisingly, the amount of SiO_2 and its molar ratios in the Alfisols and Ultisols of India are comparable with some of the Oxisols of Puerto Rico (Jones et al. 1982), Brazil (Buurman et al. 1996; Muggler

(a)   (b)

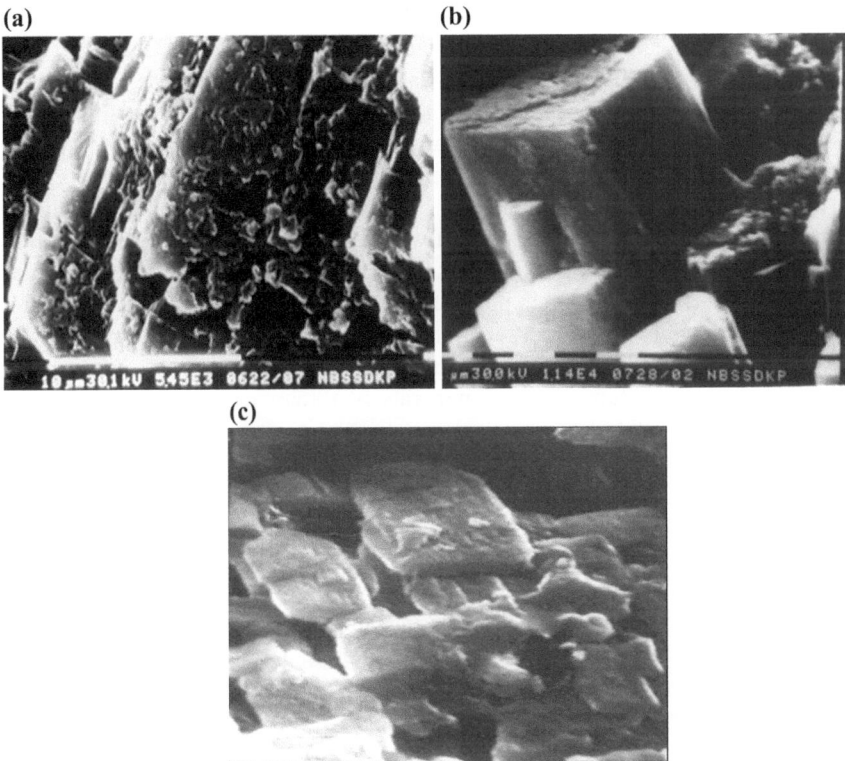

(c)

Fig. 3.7 Representative SEM photograph of sand size fractions showing gibbsite pseudomorphs after plagioclase feldspars in Ultisols of Kerala (**a**) and their hexagonal crystal shape (**b**), and highly altered sillimanite grain showing dissolution and etch pits with gibbsite crystals in Ultisols of Meghalaya (**c**) (Adapted from Chandran et al. 2005a; Bhattacharyya et al. 2000)

1998), and other regions of the World (Mohr et al. 1972). Transformation of kaolinite to gibbsite in acidic pH soils does not seem possible since kaolinite is in the intermediate weathering stage (i.e., kaolin, Kl-HIV/HIS) and not yet fully transformed to kaolinite. This particular weathering pathway suggests that gibbsite was formed in a neutral to slightly alkaline pH range and the kaolin in an acidic pH (Bhattacharyya et al. 2000).

Balasubramaniam and Sabale (1984) also suggested that kaolinisation is not an intermediate stage in the formation of bauxite. Gibbsite was formed directly from plagioclase feldspars and sillimanite in an alkaline environment. It is intriguing to note that highly acidic Alfisols and Ultisols of humid tropical Indian environment show the presence of gibbsite in soils only few states of India (Kerala, Goa, Karnataka and Meghalaya) and their KCl extractable acidity (H^+ and Al_{3+}) is much less as compared to that in other Ultisols of NEH that do not contain gibbsite (Table 3.2). Thus, a new research initiative to resolve this enigma is merited.

Alfisols and Ultisols of the HT climate contain biotite particles which survived even in kaolin dominated RF soils (Bhattacharyya et al. 2000; Chandran et al. 2005a; Pal et al. 1989). Under both optical and electron microscope examination, no gibbsite particles were observed at the fringes of biotite particles. Instead, the biotites show the initiation of layer separation due to the formation of vermiculite (Fig. 3.6a). During the earlier weathering in alkaline environment, biotites were not lost totally through their dissolution (Pal 1985) but feldspars/sillimanite and biotites were dissolved during the initial stage of weathering to produce Al in soil solution, which then crystallized as gibbsite on solid mineral surfaces. Observations under SEM on the dissolution and etch pits of feldspar/sillimanite indicated the formation of gibbsite primarily from feldspar/sillimanite (Fig. 3.7a–c) (Bhattacharyya et al. 2000; Chandran et al. 2005a). These results confirm the formation of gibbsite at the expense of primary aluminosilicate minerals is a common phenomenon in soils (Hsu 1989; Lowe 1986).

Soil pH became acidic during progressive weathering in HT climate and the released Al ions as $AL(OH)_2^+$ ions were trapped in the interlayer spaces of vermiculites/smectites to neutralize their negative charge, and to transform them into HIVs/HISs. The presence of HIV alongside Kl-HIV and M-HIV in Ultisols of Kerala, Karnataka, Meghalaya and Alfisols of Goa confirms that following the formation of gibbsite during the initial alkaline weathering, soils later experienced acidic weathering to form HIV, and finally to Kl-HIV (kaolin). Understanding the different pathways of formation of kaolinite and gibbsite, it is highly suggestive that the formation of gibbsite is not a contemporary pedogenic event in the continuing acidic humid tropical weathering environment. Thus, it would not be prudent to consider its presence in soils as an index mineral of extreme weathering conditions of soils.

3.6.2   Mixed Mineralogy Class for RF Soils of HT Climate: A Rational Proposal

Many Alfisols and Ultisols of Kerala, Karnataka and Tamil Nadu have been assigned the kaolinitic mineralogy class (Bhattacharyya et al. 2009) based on their clay CEC and ECEC values which are less than 16 and 12 cmol $(p+)kg^{-1}$, respectively (Smith 1986). It is now well understood that the prevailing acid weathering causes hydroxy interlayering of vermiculites/smectites of these soils and it is quite evident from the fact that the acidity of soils determined by $BaCl_2$-TEA is much higher than that determined by using 1 N KCl. This total acidity plus the sum of bases by NH_4OAc (pH 7) (clay CEC of sum of cations in soil control section) indicates a value much greater than 12 (Chandran et al. 2005a). Semi-quantitative estimates based on the XRD intensity of gibbsite in the <2 mm fraction and also in soil control section of few bench mark Ultisols of Kerala suggest their mineralogy belonging to either gibbsitic/allitic mineralogy class as per the US Soil Taxonomy. However, in view of the genesis of gibbsite and kaolin in Ultisols of Kerala (Chandran et al. 2005a) and

Meghalaya (Bhattacharyya et al. 2000), it would be judicious not to consider the gibbsitic/allitic mineralogy class for Ultisols and Oxisols because such mineralogy class would undermine the contemporary pedogenesis of the formation of various hydroxy-interlayered clay minerals. Therefore, the most appropriate mineralogy class for Ultisols should be 'mixed' as this class is compatible with their current land use for horticultural, forestry and agricultural crops.

3.7 Ultisols of Indian Tropical Humid Climate in Context with Doubtful Existence of Oxisols

The term 'laterite' is equivalent to Oxisols in the US Soil Classification System (Buol and Eswaran 2000). The primary requirement for Oxisols is the oxic horizon with low CEC, low ECEC, and less than 10% weatherable minerals. Oxisols are formed in HT climate, stable landscape, and siliceous/acidic parent material. Many geographical areas in India such as in Kerala, and parts of Karnataka and Tamil Nadu in southern India, Goa in the western India, the Western Ghats of India, the NEH and the Andaman and Nicobar Islands that fulfil all the requirements for the formation of Oxisols. Despite such favourable conditions the soils of these geographical areas under HT climate have not reached the stage of Oxisols (Bhattacharyya et al. 1993, 2009). Instead, the soils have kaolin (KI-HIV/HIS) and other hydroxy-interlayered minerals, and represent highly acidic Ultisols, and mildly acidic to neutral Alfisols and Mollisols. Therefore, a further critique of the published literature on Oxisols will be of much help to the world pedologists.

Beinroth (1982) reported a high amount of extractable acidity for the Puerto Rico Oxisols, which is not in accordance with the dominant presence of gibbsite and kaolinite (Jones et al. 1982) because the extractable acidity is related to the hydroxy-interlayered 2:1 layer silicate minerals. The report on the absence of hydroxy-interlayered minerals appears to be in error because such soils did show the adsorption of moderate amount of added K (Fox 1982). It is quite likely that the Puerto Rico Oxisols contain small to moderate amounts of vermiculite and/or HIV, which are responsible for the K adsorption. Macedo and Bryant (1987) reported the occurrence of Oxisols in Brazil. However, a further examination of chemical data indicates that the basic requirements of an oxic horizon are not fulfilled because these soils have clay CEC > 16 cmol (p+) kg^{-1}. Thus, the placement of these soils in Oxisols order is not well defined. Later Buurman et al. (1996) and Muggler (1998) reported results on pH and CEC of selected Oxisols, which range from 5.0 to 5.5 and 4 to 6 cmol (p+) kg^{-1}, respectively. Surprisingly, the exchangeable bases and extractable Al in most of these soils (except the surface horizons) were reported as zero. These results suggest a zero value for the ECEC and base saturation. Such a chemical environment is paradoxical, especially when the soils are still siliceous (containing still 30–50% SiO_2) and support maize cultivation. Thus, this critique brings out that the placement of these soils in Oxisols order is hardly justified.

References

Aleva GJJ (1994) Laterites: concepts, geology, morphology and chemistry. In: Creutzberg D (ed) International Soil Reference and Information Centre (ISRIC), Wageningen, The Netherlands

Balasubramaniam KS, Sabale SG (1984) Mineralogy, geochemistry and genesis of certain bauxite profiles from Kutch district, Gujarat. Proceedings of symposium on Deccan trap and bauxite. Special publication. Geological Survey of India 14, pp 225–242

Beinroth FH (1982) Some highly weathered soils of Puerto Rico. 1. Morphology, formation and classification. Geoderma 27:1–27

Bhattacharyya T, Pal DK, Deshpande SB (1993) Genesis and transformation of minerals in the formation of red (Alfisols) and black (Inceptisols and Vertisols) soils on Deccan Basalt in the Western Ghats, India. J Soil Sci 44:159–171

Bhattacharyya T, Sen TK, Singh RS, Nayak DC, Sehgal JL (1994) Morphology and classification of Ultisols with Kandic horizon in north eastern region. J Indian Soc Soil Sci 42:301–306

Bhattacharyya T, Pal DK, Deshpande SB (1997) On kaolinitic and mixed mineralogy classes of shrink-swell soils. Aust J Soil Res 35:1245–1252

Bhattacharyya T, Pal DK, Srivastava P (2000) Formation of gibbsite in presence of 2:1 minerals: an example from Ultisols of northeast India. Clay Miner 35:827–840

Bhattacharyya T, Sarkar D, Dubey PN, Ray SK, Gangopadhyay SK, Baruah U, Sehgal J, Ram Babu, Sarkar D, Mandal C, Nagar AP (2004) Soil series of Tripura. NBSS publication no. 111. NBSS&LUP, Nagpur (115 pp)

Bhattacharyya T, Pal DK, Chandran P, Ray SK (2005) Land-use, clay mineral type and organic carbon content in two Mollisols–Alfisols–Vertisols catenary sequences of tropical India. Clay Res 24:105–122

Bhattacharyya T, Pal DK, Lal S, Chandran P, Ray SK (2006) Formation and persistence of Mollisols on zeolitic Deccan basalt of humid tropical India. Geoderma 136:609–620

Bhattacharyya T, Sarkar D, Sehgal JL, Velayutham M, Gajbhiye KS, Nagar AP, Nimkhedkar SS (2009) Soil taxonomic database of India and the states (1:250, 000 scale), NBSSLUP Publ. 143, NBSS&LUP, Nagpur, India, (266 pp)

Bhattacharyya T, Pal DK, Chandran P, Ray SK, Sarkar D, Mandal C, Telpande B (2013) The clay mineral maps of Tripura and their application in land use planning. Clay Res 32:147–158

Bockheim JG, Hartemink AE (2013) Distribution and classification of soils with clay enriched horizons in the USA. Geoderma 209–210:153–160

Brunner H (1970) Pleistozäne Klimaschwankungen im Bereich des Östlichen Mysore Plateaus (Stüd Indien). Geologie 19:72–82

Buol SW, Eswaran H (2000) Oxisols. Adv Agron 68:151–195

Buurman P, Van Lagen B, Velthorst EJ (1996) Manual of soil and water analysis. Backhuys Publishers, Leiden

Chandran P, Ray SK, Bhattacharyya T, Krishnan P, Pal DK (2000) Clay minerals in two ferruginous soils of southern India. Clay Res 19:77–85

Chandran P, Ray SK, Bhattacharyya T, Dubey PN, Pal DK, Krishnan P (2004) Chemical and mineralogical characteristics of ferruginous soils of Goa. Clay Res 23:51–64

Chandran P, Ray SK, Bhattacharyya T, Srivastava P, Krishnan P, Pal DK (2005a) Lateritic soils of Kerala, India: their mineralogy, genesis and taxonomy. Aust J Soil Res 43:839–852

Chandran P, Ray SK, Bhattacharyya T, Pal DK (2005b) Chemical and mineralogical properties of ferruginous soils of Andaman and Nicobar Islands. Abstract, 70th annual convention and national seminar on "Developments of soil science" of the Indian Society of Soil Science. TNAU, Coimbatore, Tamil Nadu, p 45

Chandran P, Ray SK, Bhattacharyya T, Sen TK., Sarkar D, Pal DK (2006) Rationale for mineralogy class of ferruginous soils of India. Abstract, 15th annual convention and national symposium on "Clay research in relation to agriculture, environment and forestry" of the Clay Minerals Society of India. BCKVV, Mohanpur, West Bengal, p 1

Das AL, Goswami A, Thampi CJ, Sarkar D, Sehgal J (1996) Soils of Andaman and Nicobar Islands for optimizing land use, NBSS Publ. 61 (Soils of India Series), National Bureau of Soil Survey and Land Use Planning, Nagpur, India, 57 pp + 5 sheets of soil map (1:50,000 scale)

Eswaran H (1972) Micromorphological indicators of pedogenesis in some tropical soils derived from basalts from Nicaragua. Geoderma 7:15–31

Eswaran H, Sys C (1979) Argillic horizon in LAC soils formation and significance to classification. Pédologie 29:175–190

Eswaran H, Kimble J, Cook T, Beinroth FH (1992a) Soil diversity in the tropics: implications for agricultural development. In: Lal R, Sanchez PA (eds) Myths and science of soils of the tropics. SSSA special publication number 29. SSSA, Inc. and ACA, Inc., Madison, Wisconsin, USA, pp 1–16

Eswaran H, Krishnan P, Reddy RS, Reddy PSA, Sarma VAK (1992b) Application of 'Kandi' concept to soils of India. J Indian Soc Soil Sci 40:137–142

Fox RL (1982) Some highly weathered soils of Puerto Rico, 2. Chemical properties. Geoderma 27:139–176

Gowaikar AS (1973) Influence of moisture regime on the genesis of laterite soils in south India. III. Soil classification. J Indian Soc Soil Sci 21:343–347

Harindranath CS, Venugopal KR, Raghu Mohan NG, Sehgal J, Velayutham MV (1999) Soils of Goa for optimising land use. NBSS Publ. 74b (Soils of India Series), National Bureau of Soil Survey and Land Use Planning, Nagpur, India, 131 pp + 2 sheets of soil map on 1:500,000 scale

Hsu PH (1989) Aluminium hydroxides and oxyhydroxides. In: Dixon JB, Weed SB (eds) Minerals in soil environments. Soil Science Society of America, Madison, WI, pp 331–378

Jackson ML (1964) Chemical composition of soils. In: Bear FE (ed) Chemistry of the soil. Van Norshtand-Reenhold, New York, pp 71–141

Jones RC, Hundall WH, Sakai WS (1982) Some highly weathered soils of Puerto Rico, 3. Mineralogy. Geoderma 27:75–137. doi:10.1016/0016-7061(82)90048-9

Kharche VK (1996) Developing soil-site suitability criteria for some tropical plantation crops. Ph. D thesis, Dr. P D K V, Akola, Maharashtra, India

Kooistra MJ (1982) Micromorphological analysis and characterization of 70 benchmark soils of India. Soil Survey Institute, Wageningen, The Netherlands

Krishnan P, Venugopal KR, Sehgal J (1996) Soil resources of Kerala for land use planning. NBSS Publ. 48b (Soils of India Series 10), National Bureau of Soil Survey and Land Use Planning, Nagpur, India, 54 pp + 2 sheets of soil map on 1:500,000 scale

Lal S, Deshpande SB, Sehgal JL (1994) Soil series of India. Soils bulletin no. 40. National Bureau of Soil Survey and Land Use Planning (ICAR), Nagpur, India, (684 pp.)

Lowe DL (1986) Controls on the rate of weathering and clay mineral genesis in airfall tephras: a review and New Zealand case study. In: Colman SM, Dethier DP (eds) Rates of chemical weathering of rocks and minerals. Academic Press, Orlando, FL, pp 265–330

Macedo J, Bryant RB (1987) Morphology, mineralogy and genesis of a hydro sequence of Oxisols in Brazil. Soil Sci Soc Am J 51:690–698

Millot G (1970) Geology of clays. Springer-Verlag, New York

Mohr ECJ, Van Baren FA, van Schuylenborgh J (1972) Tropical soils-a comprehensive study of their genesis. Mouton, The Hague, The Netherlands

Muggler CC (1998) Polygenetic Oxisols on tertiary surfaces, Minas Gerais, Brazil: soil genesis and landscape development. (PhD thesis) Wageningen Agricultural University, The Netherlands

Murthy RS, Hirekerur LR, Deshpande SB, Venkat Rao BV (eds) (1982) Benchmark soils of India. National Bureau of Soil Survey and Land Use Planning, Nagpur, India (374 pp)

Natarajan A, Reddy PSA, Sehgal J, Velayutham M (1997) Soil resources of Tamil Nadu for land use planning. NBSS publ. 46b (Soils of India Series), National Bureau of Soil Survey and Land Use Planning, Nagpur, India, 88 pp + 4 sheets of soil map on 1:500,000 scale

Nayak DC, Sen TK, Chamuah GS, Sehgal JL (1996) Nature of soil acidity in some soils of Manipur. J Indian Soc Soil Sci 44:209–214

Pal DK (1985) Potassium release from muscovite and biotite under alkaline conditions. Pedologie (Ghent) 35:133–146

Pal DK, Deshpande SB (1987) Genesis of clay minerals in a red and black complex soil of southern India. Clay Res 6:6–13

Pal DK, Deshpande SB, Durge SL (1987) Weathering of biotite in some alluvial soils of different agro climatic zones. Clay Res 6:69–75

Pal DK, Deshpande SB, Venugopal KR, Kalbande AR (1989) Formation of di- and trioctahedral smectite as an evidence for paleoclimatic changes in southern and central Peninsular India. Geoderma 45:175–184

Pal DK, Kalbande AR, Deshpande SB, Sehgal JL (1994) Evidence of clay illuviation in sodic soils of north-western part of the Indo-Gangetic plains since the Holocene. Soil Sci 158:465–473

Pal DK, Dasog GS, Vadivelu S, Ahuja RL, Bhattacharyya T (2000a) Secondary calcium carbonate in soils of arid and semi-arid regions of India. In: Lal R, Kimble JM, Eswaran H, Stewart BA (eds) Global climate change and pedogenic carbonates. Lewis Publishers, Boca Raton, Florida, pp 149–185

Pal DK, Bhattacharyya T, Deshpande SB, Sarma VAK, Velayutham M (2000a) Significance of minerals in soil environment of India. NBSS Review Series 1. NBSS&LUP, Nagpur, (68 pp)

Pal DK, Sarma VAK (2002) Chemical composition of soils. Fundamentals of soil science. Indian Society of Soil Science, New Delhi, pp 209–227

Pal DK, Srivastava P, Bhattacharyya T (2003) Clay illuviation in calcareous soils of the semi-arid part of the Indo-Gangetic Plains, India. Geoderma 115:177–192

Pal DK, Bhattacharyya T, Chandran P, Ray SK, Satyavathi PLA, Durge SL, Raja P, Maurya UK (2009) Vertisols (cracking clay soils) in a climosequence of Peninsular India: evidence for Holocene climate changes. Quatern Int 209:6–21

Pal DK, Wani SP, Sahrawat KL (2012) Vertisols of tropical Indian environments: pedology and edaphology. Geoderma 189–190:28–49

Pal DK, Sarkar D, Bhattacharyya T, Datta SC, Chandran P, Ray SK (2013) Impact of climate change in soils of semi-arid tropics (SAT). In: Bhattacharyya T, Pal DK, Sarkar D, Wani SP (eds) Climate change and agriculture. Studium Press, New Delhi, pp 113–121

Pal DK, Wani SP, Sahrawat KL, Srivastava P (2014) Red ferruginous soils of tropical Indian environments: A review of the pedogenic processes and its implications for edaphology. Catena: 121:260–278. DOI:10.1016/j. catena 2014.05.023

Ray SK, Chandran P, Durge SL (2001) Soil taxonomic rationale: kaolinitic and mixed mineralogy classes of highly weathered ferruginous soils. Abstract, 66th annual convention and national seminar on "Developments in Soil Science" of the Indian Society of Soil Science, Udaipur, Rajasthan, pp 243–244

Rebertus RA, Buol SW (1985) Intermittency of illuviation in Dystrochrepts and Hapludults from the Piedmont and Blue Ridge province of North Carolina. Geoderma 36:277–291

Rengasamy P, Krishna Murthy GSR, Sarma VAK (1975) Isomorphous substitution of iron for aluminium in some soil kaolinites. Clays Clay Miner 23:211–214

Rengasamy P, Sarma VAK, Murthy RS, Krishna Murthy GSR (1978) Mineralogy, genesis and classification of ferruginous soils of the eastern Mysore Plateau, India. J Soil Sci 29:431–445

Sanchez PA (1976) Properties and management of soils in the tropics. Wiley, New York

Sanchez PA, Logan TJ (1992) Myths and science about the chemistry and fertility of soils. SSSA special publication number 29. In: Lal R, Sanchez PA (eds) Myths and science of soils of the tropics. SSSA, Inc and ACA, Inc, Madison, Wisconsin, USA, pp 35–46

Schwertmann U, Herbillon AJ (1992) Some aspects of fertility associated with the mineralogy of highly weathered tropical soils. In: Lal R, Sanchez PA (eds) Myths and science of soils of the tropics. SSSA special publication number 29. SSSA, Inc and ACA, Inc, Madison, Wisconsin, USA, pp 47–59

Sehgal JL (1998) Red and lateritic soils: an overview. In: Sehgal J, Blum WE, Gajbhiye KS (eds) Red and lateritic soils. Managing red and lateritic soils for sustainable agriculture. Vol. 1. Oxford and IBH Publishing Co. Pvt. Ltd., New Delhi, pp 3–10

Sehgal JL, Challa O, Thampi CJ, Maji AK, Naga Bhusana SR (1998) Red and lateritic soils of India. In: Sehgal J, Blum WE, Gajbhiye KS (eds) Red and lateritic soils. Managing red and lateritic soils for sustainable agriculture, vol 2. Oxford and IBH Publishing Co. Pvt. Ltd., New Delhi, pp 1–10

Sen TK, Chamuah GS, Sehgal JL (1994) Occurrence and characteristics of some Kandi soils of Manipur. J Indian Soc Soil Sci 42:297–300

Sen TK, Dubey PN, Chamuah GS, Sehgal JL (1997a) Landscape–soil relationship on a transect in central Assam. J Indian Soc Soil Sci 45:136–141

Sen TK, Nayak DC, Singh RS, Dubey PN, Maji AK, Chamuah GS, Sehgal JL (1997b) Pedology and edaphology of benchmark acid soils of north-eastern India. J Indian Soc Soil Sci 45:782–790

Sen TK, Nayak DC, Dubey PN, Chamuah GS, Sehgal JL (1997c) Chemical and electrochemical characterization of some acid soils of Assam. J Indian Soc Soil Sci 45:245–249

Shiva Prasad CR, Reddy PSA, Sehgal J, Velayutham M (1998) Soils of Karnataka for optimising land use. NBSS Publ. 47b (Soils of India Series), National Bureau of Soil Survey and Land Use Planning, Nagpur, India, 111 pp + 4 sheets of soil map on 1: 500,000 scale

Smith GD (1986) The Guy Smith interviews: rationale for concept in soil taxonomy. SMSS technical monograph, 11. SMSS, SCS, USDA, USA

Srivastava P, Chandran P, Ray SK, Bhattacharyya T (2001) Evidence of chemical degradation in tropical ferruginous soils of southern India. Clay Res 20:31–41

Srivastava P, Bhattacharyya T, Pal DK (2002) Significance of the formation of calcium carbonate minerals in the pedogenesis and management of cracking clay soils (Vertisols) of India. Clays Clay Miner 50:111–126

Srivastava P, Rajak M, Sinha R, Pal DK, Bhattacharyya T (2010) A high resolution micromorphological record of the late quaternary Paleosols from Ganga-Yamuna Interfluve: stratigraphic and paleoclimatic implications. Quatern Int 227:127–142

Soil Survey Staff (1990) Keys to soil taxonomy, SMSS technical monograph fourth edition, 19. Blacksburg, Virginia, USA

Tait JM, Violante A, Violante P (1983) Co-precipitation of gibbsite and bayerite with nordstrandite. Clay Miner 18:95–97

Van Olphen H (1966) An introduction of clay colloid chemistry. Interscience, New York

Varghese T, Byju G (1993) Laterite soils, technical monograph no. 1. State committee on science, technology and environment. Government of Kerala, Kerala, India

Venugopal KR (1997) Types of cutans in some ferruginous soils of Bangalore plateau and their relation with soil development. J Indian Soc Soil Sci 46:641–646

Venugopal KR, Deshpande SB, Kalbande AR, Sehgal JL (1991) Textural pedofeatures (clay coatings) in a ferruginous soil from Bangalore plateau. Clay Res 10:30–35

Chapter 4
Soils of the Indo-Gangetic Alluvial Plains: Historical Perspective, Soil-Geomorphology and Pedology in Response Climate Change and Neotectonics

Abstract Indian earth scientists and soil scientists based on large number of well-presented pedons in the Indo-Gangetic Alluvial Plains (IGP), spread along the west hot arid climate to per-humid climate in the east, have led to new perspectives on the historical development of the IGP and the soils therein. This addresses the hitherto little known subtleties of pedogenesis and polygenesis due to recorded tectonic, climatic and geomorphic episodes and phenomena during the Holocene. Based on degree of development, five geomorphic surfaces, with soil ages 0.5, 0.5–2.5, 2.5–5.0, 5.0–10, >10 ka respectively, are mappable in the IGP and correspond to the post-incisive chronosequences that evolved in response to interplay of fluvial processes, climatic fluctuations, and neotectonics during the Holocene. The polygenetic signatures, illuvial clay pedofeatures, pedogenic carbonates, clay mineralogy, and stable isotope geochemistry, suggest the evolution of the IGP soils witnessed two humid phases (13.5–11.0 and 6.5–4.0 ka) with intervening dry climatic conditions. The IGP soils with varying climate from hot-arid to per humid belong to Entisols, Inceptisols, Alfisols, and Vertisols orders. Addition and depletion of OC, formation pedogenic $CaCO_3$, illuviation of clay particles and argilli-pedoturbation are the major pedogenic processes in soils of the IGP during the Holocene. The IGP soils are, in general, micaceous, but the soils with vertic characters are smectitic. A better understanding of the pedology of the IGP soils and their linkage to climate change, and landscape stability appear to be potentially useful as guideline for their management. Thus the new knowledge base has potential as a reference for critical assessment of the pedosphere for health and quality in different parts of the world and may facilitate developing a suitable management practices for the food security in the 21st century.

Keywords Indo-Gangetic plains (IGP) · Soil-geomorphology · Climate change · Neotectonics · Polygenesis

© Springer International Publishing AG 2017   71
D.K. Pal, *A Treatise of Indian and Tropical Soils*,
DOI 10.1007/978-3-319-49439-5_4

4.1 Introduction

The Indo-Gangetic Plains (IGP) ranks as one of the most extensive fluvial plains in the world. The deposit of this tract represents the last chapter of earth's history. This plain is bound by the Himalaya in the north and Craton to the south (Fig. 4.1). Geological and geophysical investigations indicate that it is a vast asymmetric

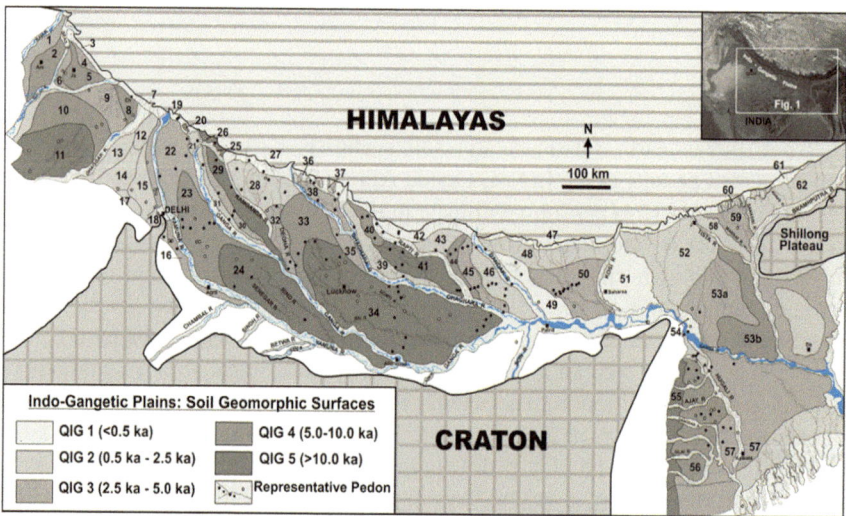

Fig. 4.1 Soil-geomorphic map of the IGP extending from Ravi River in the west to the Deltaic Plains in the east. The 62 soil-geomorphic units identified in the IGP stand for: *1* Ravi-Beas interfluve I, *2* Ravi-Beas interfluve II, *3* Sutluj-Beas piedmont, *4* Sutluj-Beas interfluve I, *5* Sutluj-Beas interfluve II, *6* Sutluj-Beas interfluve III, *7* Sutluj-Ghagghar piedmont I, *8* Sutluj-Ghagghar piedmont II, *9* Sutluj-Ghagghar interfluve I, *10* Sutluj-Ghagghar interfluve II, *11* Sutluj-Ghagghar interfluve III, *12* Ghagghar-Yamuna interfluve I, *13* Ghagghar-Yamuna interfluve II, *14* Ghagghar-Yamuna interfluve III, *15* Ghagghar-Yamuna interfluve IV, *16* Old Yamuna plains, *17* Aeolian plains I, *18* Aeolian plains II, *19* Young Piedmont plains, *20* Old Piedmont plains I, *21* Old Piedmont plains II, *22* Yamuna-Ganga interfluve I, *23* Yamuna-Ganga interfluve II, *24* Yamuna-Ganga interfluve III, *25* Old Piedmont plain II, *26* Old Piedmont plain I, *27* Young Piedmont plain, *28* Kosi-Gola plains, *29* Ganga-Ramganga interfluve, *30* Old Ganga plains, *31* Young Ganga plains, *32* Deoha plains, *33* Ganga-Ghaghara interfluve I, *34* Ganga-Ghaghara interfluve II, *35* Old Ghaghara plains I, *36* Young Sihali-Kandra piedmont, *37* Old Sihali-andra piedmont, *38* Old Ghaghara plains II, *39* Young Ghaghara plains, *40* Rapti-Ghaghara interfluve I, *41* Rapti-Ghaghara interfluve II, *42* Gholia-Dhobania piedmont, *43* Old Rapti plains, *44* Oldest Gandak plain, *45* Older Gandak plain, *46* Young Gandak plain, *47* Gandak-Kosi piedmont I, *48* Gandak-Kosi piedmont II, *49* Gandak-Kosi interfan I, *50* Gandak-Kosi interfan II, *51* Kosi Megafan, *52* Mahananda-Tista plains, *53a* Barind Tract 1, *53b* Barind Tract II, *54* Old Ganga Plain, *55* Red soil uplands, *56* Ajay-Damodar-Silai plains, *57* Hugali-Ganga plains, *58* Tista-Mansai plains, *59* Mansai-Sankosi plains, *60* Sankosi-Manas Piedmont I, *61* Sankosi-Manas piedmont II, *62* Sankosi-Manas piedmont I. The soil profiles marked as *solid circles* represent pedons from the earlier and recent studies and the open circles represent new data (Adapted from Srivastava et al. 2015)

trough with maximum thickness of about 10 km in the north and a minimum of few meters in the south towards Craton (Wadia 1966; Sastri et al. 1971; Rao 1973; Raiverman et al. 1983; Parkash and Kumar 1991). Origin of the IGP is related to the Himalayan orogeny caused by the collision of Indian and Tibetan plates at ca 50 Ma (Klootwijk et al. 1992; Gaina et al. 2007; Kumar et al. 2007; Kent and Muttoni 2008). The Indian Plate is still moving at the rate of 2–5 cm/y towards the north and the compression generated throughout the plate ensures that it is continuously under stress providing the basic source of strain in the fractured zones (Gaur 1994). In response to continued northward push of the Indian Plate fluvial deposits, landforms, and soils of the IGP are influenced by the neotectonics during the Holocene (Srivastava et al. 1994; Parkash et al. 2000). The major rivers of the IGP have changed their courses and, at present, are flowing in southeast and easterly directions with convexity towards the southeast, which is strikingly similar to the arcuate pattern of the major thrusts bordering the IGP (Parkash et al. 2000). The IGP is remarkably flat with <0.02% topographic gradient towards east from Panjab in the west ($\sim$300 m amsl) to the deltaic region (<10 m amsl) in the east. The IGP is mainly drained by the Himalayan Rivers, the Indus, the Yamuna, the Ganga, the Ghaghara, and the Brahmaputra rivers and also by the rivers originating from the south in the Cratonic region, the Chambal, the Betwa, and the Son Rivers.

The surface of the IGP is covered with fluvial sediments and soils with varying characteristics over an aerial extent of about 52.01 m ha and represents 17 agro-ecological sub-regions (AESRs) (Mandal et al. 2014). The IGP accounts for the one third cultivable lands and a major contributor of the food supply for more than 200 million population of India. The evidence from many archaeological sites of the IGP show considerable changes from incipient agricultural activities to well-developed agricultural practices in this plain during the last 10 ka (Sharma et al. 1980; Williams et al. 2006; Kumar et al. 1996; Misra 2001; Saxena et al. 2006; Fuller 2006). Agriculture was the mainstay of the people of ancient India and the agriculturists were aware of different types of soils and their productivities (Velayutham and Pal 2004). Archaeological evidences further indicate that western parts of the IGP were the ancient agricultural hub ($\sim$10 ka) from where dispersal of domestication and anthropogenic activities took place towards the eastern and the southern parts of India (Fuller 2006; Chen et al. 2010). In the Indus valley, the earliest record of cultivation from northern Gujrat is placed at 3500 BC (Patel 1999; Meadow and Patel 2003). In the Gangetic Plains, the earliest agriculture dates back to early Harrapan period (2800 BC) with both summer and winter crops (e.g. site Kunaal; Fuller 2006), suggesting that the monsoonal crops were already available for cultivation over western parts of the IGP during the Harrapan period. This agricultural practice then spread further east into the Gangetic Plains. The period after 2000 BC is marked by agricultural village settlement over a wider region, covering landscapes with agriculturists and sedentary settlements (Fuller 2006).

4.2   Early Studies on Soils

The surface soil remains the firm foundation of human life thriving on it. However, anthropogenic intervention over the last 10 ka has made it one of the most degraded and least understood ecosystems in terms of its relationship with people. The knowledge of the IGP soils is important for understanding the long-term human interactions (Williams and Clarke 1984, 1995; Williams et al. 2006; Gibling et al. 2008). The ancient literature of India ('Vishnu Purana' from the first century C.E.) suggests that the early farmers had the knowledge of soils, landforms, erosion, flooding, sedimentation, vegetation, land use, water and human health (McNeill and Winiwarter 2004; Wasson 2006). Scientific characterization the IGP soils can be traced back to the first half of 19th century when the Geological Survey of India started studying the soils and the underlying strata in 1846 (Velayutham et al. 2002). It was observed that soil group boundaries were nearly co-incident with the major litho-tectonic boundaries (Wadia et al. 1935). The IGP soils were defined as one of the four major soils groups in India (Voelcker 1893; Leather 1898). A synthesis of pedological investigations carried out over the last several decades indicates various soil-forming processes such as calcification, leaching, lessivage, salinization-alkalinisation, gleization and homogenization in soils of the IGP (Shankarnarayana and Sarma 1982). These processes lead to the formation of a variety of soils in the IGP represented mainly by three soil orders i.e. Entisols, Inceptisols, and Alfisols (Shankarnarayana and Sarma 1982). Recent studies, however, indicate that Mollisols, Aridisols, and Vertisols are also present in the IGP (Bhattacharyya et al. 2004; Ray et al. 2006). The revised soil map of the IGP depicts the presence of newly added soil orders (Mandal et al. 2014).

4.3   Soil-Geomorphology of the Indo-Gangetic Plains (IGP)

Over the last three decades pedological and geomorphic methods were applied in Indian soils to decipher the soil-geomorphic history in different parts of the IGP (Singhai et al. 1991; Mohindra, et al. 1992; Srivastava et al. 1994; Kumar et al. 1996; Singh et al. 1998; Khan et al. 2005). Despite the significant pedological information of the IGP soils from different parts now available, a comprehensive understanding of the entire IGP in terms of pedogenic response to climate change, and neotectonics remains a major gap to be attended by the researchers (Srivastava et al. 2015). This treatise on the IGP soils is to provide a critical synthesis of research data available on the soils in the IGP for understanding the pedogenic response to climate change and neotectonics during the Holocene period, and it will be a relevant reference to the researchers around the world on account of unique position of the Indian subcontinent and the rising Himalayas that influence monsoon, landscapes and vegetation at both regional and global scales (Srivastava et al. 2015). In addition, it will help establish the contemporary pedogenic processes as well as polygenetic pedogenic history from different geomorphic surfaces, which are so necessary for optimizing and sustaining their efficient use and management under national agricultural research systems (NARS).

The application of soil-geomorphic approach is well developed technique, which helps in establishing the genetic relationships between the soils and the landscape elements (McFadden and Knuepfer 1990; Birkeland 1990). The approach is based on the fact that soils form an essential component of any landscape and the history of any landscape evolution is intimately tied with the history of soil development (Birkeland 1990). The application of a soil geomorphic approach to soil landscape studies has led to a more quantitative evaluation of the geomorphic processes operating over different time scales ranging from 10^3 to 10^5 years (Ritter 1986). Soil-geomorphic studies over the last three decades have demonstrated that soil development 'functions' help in determining the interrelationship among time, climate, landscape development and the soil-forming processes (Birkeland 1990; Harden 1990; McFadden and Knuepfer 1990).

Soil-geomorphological research in India, started in the 1980s that covered almost 40% of the IGP as foundation to soil geomorphic investigations (Singhai et al. 1991; Mohindra, et al. 1992; Srivastava et al. 1994; Kumar et al. 1996; Singh et al. 1998; Khan et al. 2005). An integrated approach involving pedological and geomorphic methods was used as tools to unravel Holocene soil-geomorphic history in different parts of the IGP. The research work carried out has established that the IGP mainly characterized by more than two geomorphic surfaces with polygenetic soils on older geomorphic surfaces that developed in response to climate change and neotectonics over the last 13.5 ka (Srivastava et al. 1994; Srivastava and Parkash 2002).

Recent advances on soil-geomorphic research in the entire IGP extending from the Ravi River in the west to deltaic plains in the east are now available (Srivastava et al. 2015). A brief summary of this indicates that soil-geomorphology of the IGP is described in terms of soil-geomorphic units; degree of soil development, geomorphic surfaces, and soil-geomorphic groups. It is based on the published soil-geomorphic work and recent research on large number of new pedons (Fig. 4.1; Srivastava et al. 2015). The entire IGP from the arid west to the per-humid east comprises of 62 soil geomorphic units and 12 soil-geomorphic groups that are spread over different landforms like floodplains, piedmonts, alluvial plains, old river plains, and interfluves (Fig. 4.1; Srivastava et al. 2015). Each of the 62 soil-geomorphic units can be identified by distinctive features on IRS images on account of varying characteristics of surface soils and sediments (Mohindra et al. 1992; Srivastava et al. 1994). The characterization of a geomorphic surface is based on degree of development and duration of pedogenesis (Mohindra et al. 1992; Srivastava et al. 1994; Singh et al. 1998). Pedogenic studies of the IGP confirm that origin and stabilization of different geomorphic surfaces occurred during slight uplift of a region above the general level of rivers that led to the termination of sedimentation and initiation of pedogenic activity (Srivastava et al. 1994; Kumar et al. 1996; Singh et al. 1998). This led to the formation of the least developed soils close to channels, alluvial fans, and piedmonts where sediment accumulation rates are high, whereas the most developed soils formed in upland interfluves and old river plains, where sediment accumulation rates are significantly less (Srivastava et al. 1994; Kumar et al. 1996; Singh et al. 1998). The mature/well-developed soils on old geomorphic surfaces with an age of 8000–13,500 years B.P exhibit strongly to very strongly developed pedofeatures. By contrast, the immature or poorly developed soils with little or no pedogenesis formed on young geomorphic surfaces over the last 500 years. The degree of soil development in each of the 62 units was critically evaluated and compared with available soil-geomorphic data of the IGP to determine the type of geomorphic surfaces. Following these criteria at least five geomorphic surfaces, namely QIG1–QIG5 with ages of <0.5, 0.5–2.5, 2.5–5.0, 5–10, and >10 ka, are delineated on the 62 soil-geomorphic units (Fig. 4.1; Srivastava et al. 2015). These five geomorphic surfaces of the IGP are similar to the geomorphic surfaces in interfluves of the Greenfield Quadrangle, Iowa (Ruhe 1969). The IGP geomorphic surfaces are comparable to the post-incisive soil chronosequences marked by a sequence of progressively younger surfaces in which soils formation may begin (Vreeken 1975; Mohindra et al. 1992; Srivastava et al. 1994; Huggett 1998). Each of the five geomorphic surfaces of the IGP is characterized by a distinctive pedogenic record and typical pedogenic features (Srivastava et al. 2015).

4.4 Soil Ages of the IGP

In the recent past the IGP was considered to consist of an older and a younger alluvium (Wadia 1966; Bhattacharya and Banerjee 1979). Marked by a higher degree of pedogenesis, the older alluvium was assigned a lower Pleistocene age (Wadia 1966; Bhattacharya and Banerjee 1979) or a minimum age of 120 ka (Singh 1988). Soil-geomorphic studies during the early to end of nineties demonstrate the presence of more than two geomorphic surfaces in the IGP (Mohindra et al. 1992; Srivastava et al. 1994; Kumar et al. 1996; Srivastava et al. 1998; Singh et al. 1998). Age of the IGP soils that formed on five geomorphic surfaces range from <0.5 to 13.5 ka (Srivastava et al. 2015). The soil ages are based on radiocarbon (C^{14}), thermoluminescence (TL), archaeological, and historical data from the IGP (Srivastava et al. 2015). The TL dates are greater than the radiocarbon dates of calcretes because TL dates provide the time of deposition and start of pedogenesis, whereas calcrete takes time to form and it is also affected by dissolution and reprecipitation within the soil profile (Srivastava et al. 1994; Singh et al. 1998). TL dates are considered here as approximate maximum ages of soils. The 'Lateritic Upland' (unit 55, Fig. 4.1; Srivastava et al. 2015) with red soils of Early to Middle Pleistocene age is the oldest soil of the IGP (Singh et al. 1998).

4.5 Pedogenic Response of the IGP Soils to Holocene Climatic Fluctuations

The soil-geomorphic research of the IGP has helped in describing the climatic changes that are in pedogenic features of the soils developed on different geomorphic surfaces (Srivastava et al. 2015). In this regard micromorphology, clay minerals, isotope geochemistry of pedogenic carbonates, and polygenesis act as important proxies for Holocene climate changes (Srivastava et al. 1998; Srivastava 2001; Srivastava and Parkash 2002). Information from these proxies indicates two humid phases during soil formation over the IGP, one between 13.5 and 11.0 ka and the other between 6.5 and 4.0 ka. The intervening period witnessed the dry climate (Srivastava et al. 1998; Srivastava 2001; Srivastava and Parkash 2002).

Soil thin section studies indicate that the Q1G5 soils on uplands of the IGP are polygenetic (Srivastava and Parkash 2002) as evidenced from the presence of degraded illuvial clay pedofeatures of an early humid phase (13.5–11.0 ka) (Fig. 4.2a) and thick (150–200 μm) micro-laminated clay pedofeatures of a later humid phase (6.5–4.0 ka). The earlier clay pedofeatures show degradation as bleaching, loss of preferred orientation, development of a coarse speckled appearance and fragmentation. Clay pedofeatures of the later phase are thick, smooth and strongly birefringent micro-laminated clay pedofeatures (Fig. 4.2b; Srivastava and Parkash 2002). The clay illuviation was more extensive during the later phase as the groundmass was enriched by discrete pedofeatures of clay intercalations (Srivastava and Parkash 2002; Srivastava et al. 2015, 2016).

Formation of PCs was dominant during the dry phases of climate (early Holocene to 6.5 ka and after 4.0 ka) while their dissolution-reprecipitation was

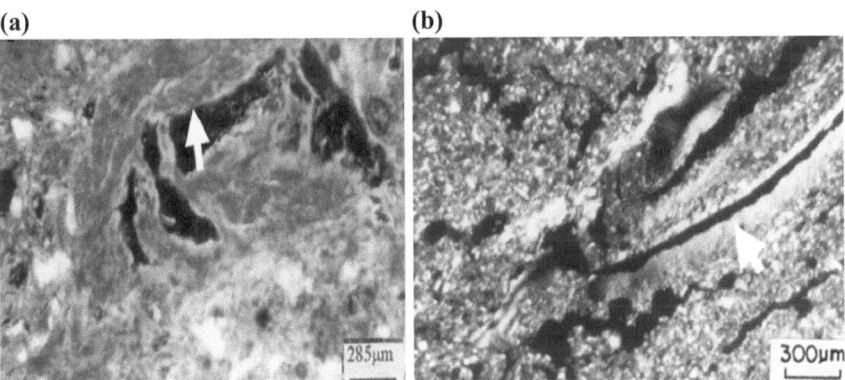

(a)   **(b)**

Fig. 4.2 Representative photomicrographs in cross-polarized light of **a** the thick, degraded, illuvial clay pedofeatures and **b** thick clay pedofeatures of Alfisols of the IGP (Adapted from Pal et al. 2009b)

noticed in the subsequent wetter climates (Srivastava 2001; Srivastava and Parkash 2002). Clay mineralogical studies of IGP soils show that biotite weathered to trioctahedral vermiculite and smectite during the semiarid-arid climate and it transformed to interstratified smectite-kaolin (Sm/K) during the warm and humid climate (Fig. 4.3; Srivastava et al. 1998). The relatively dry climates since 4.0 ka

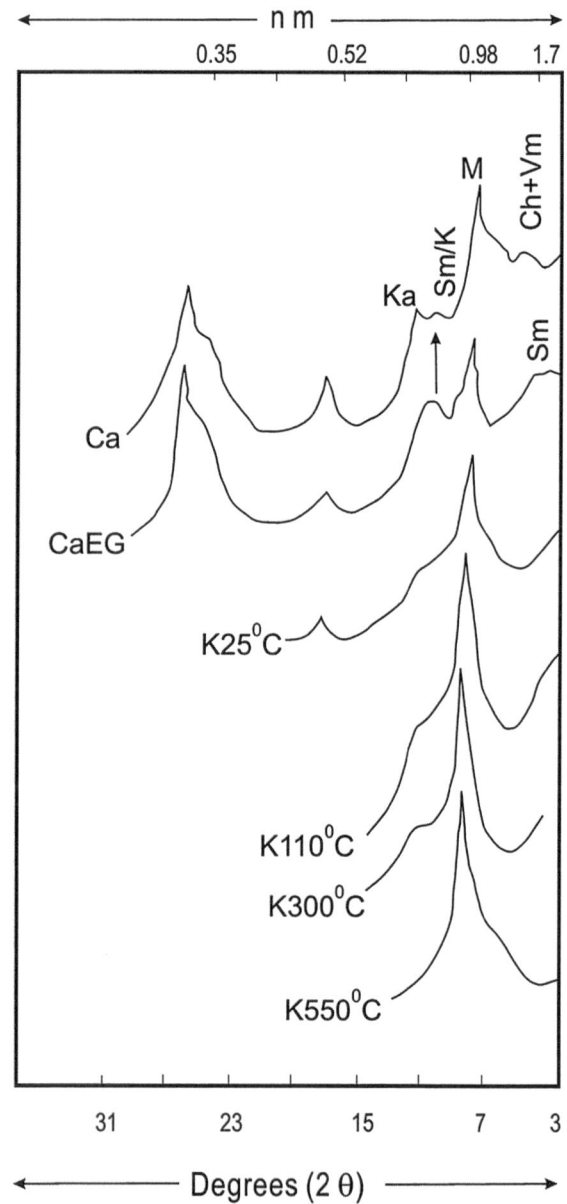

Fig. 4.3 Representative XRD diagrams of the fine clay fractions of the QGH5 soils: Ca–Ca, saturated; CaEG-Ca, saturated and glycolated: K25, K110; K300, and K5500 °C-K, saturated and heated; *S* smectite, *V* vermiculite; *Ch* chlorite; *Sm/K* smectite-kaolin; *K* kaolinite; *ML* mixed layer (1.0–1.4 nm) Adapted from Srivastava et al. 1998)

lead to the preservation of vermiculite, smectite, and Sm/K assemblages in older IGP soils (Srivastava et al. 1998). The Carbon and Oxygen isotope studies on the pedogenic carbonates from the older soils (8–10 ka) indicate a relative depletion of ^{13}C compared to younger soils (<2.5 ka), indicating low soil-respiration and drier conditions during the early Holocene (Srivastava 2001). The inferences on paleo-climatic changes made through PC of the IGP soils are compatible with those obtained by the paleovegetation record (Gupta 1978; Srivastava 2001). The Holocene paleovegetation record of the IGP indicates the spread of *Chenopodium and Typha angustata* biomass during the semiarid-arid climate that changed to *Anogeissus and Tecomella spp.* in warm and humid climatic phase (Gupta 1978).

4.6   Pedogenic Response of the IGP Soils to Neotectonic Events

The tectonic framework of the IGP is marked by basement and surface structural features as described by Srivastava et al. (2015). The surface structural features such as faults and tectonic blocks have been identified on the basis of mapping soil-geomorphic features in the IGP. Some of the surface faults are related to the basement and these have been active over the last 10 ka. These faults in the IGP are considered analogues to the growth faults in the deltaic regions (Mohindra et al. 1992; Srivastava et al. 1994; Singh et al. 1998). Development of the surface structural features in the IGP is mainly due to the movement of the Indian Plate, causing compressive stresses from the south (Srivastava et al. 1994; Parkash et al. 2000).

Neotectonics influenced the pedogenesis along the geomorphic surfaces of the IGP the during the Holocene. The pedogenic response to the neotectonics can be explained by tectonic movement of the fault bounded blocks (Srivastava et al. 1994). During the uplift of the blocks above the general level of rivers, sedimentation was terminated, which was then followed by the initiation of pedogenic activity. Soils formed on the uplifted blocks, therefore, preserve the climate record. The uplift of different blocks at different times provides a sequence of soils that preserve signatures of the climate and the extent of soil development. The older soils are also modified by the later climate showing polygenetic features (Srivastava and Parkash 2002). In view of the neotectonics of the IGP, following features within soils and geomorphic units of the IGP can be summarized in three ways. (i) Nearly 50% of the IGP is covered by upland region with well-developed soils on QIG4–QIG5 surfaces in the western part of the Ghagghar, upland interfluves of the Yamuna-Ganga, Ganga-Ghaghara, Ghaghara-Rapti interfluves and some upland regions of the deltaic plains (unit 55–56, Fig. 4.1; Srivastava et al. 2015). It implies strong pedogenic activity since ~ 10 ka over these stable upland regions and active sedimentation in the remaining parts of the IGP, which ended with poorly developed soils (Srivastava et al. 1994; Kumar et al. 1996; Singh et al. 1998).

(ii) Sagging or down warping of large geomorphic surfaces or tectonic blocks bounded by faults resulted in greater sinuosity of the streams such as, the Gomti, Sai, Rapti and Sarju rivers in units 34, 40, 41; (Fig. 4.1; after Srivastava et al. 1994). (iii) The compressive forces from SW modified the otherwise flat terrain to micro-low (ML) (0.5–1.0 m) and micro-high (MH). Such neotectonics have led to the formation of highly sodic soils (Natrustalfs/Natraqualfs) in micro-low and less sodic/non-sodic soils on micro-high areas of QIG4–QIG5 geomorphic surfaces (Fig. 4.1) in parts of the Ganga-Yamuna interfluve (Pal et al. 2003b) (pedogenesis of these soils are detailed in Chap. 6).

Micromorphological thin section studies of MH and ML soils show deformational pedofeatures such as cross and reticulate striation of plasmic fabric (Fig. 4.4a), disruption of clay pedofeatures (Fig. 4.4b), and carbonate nodules (Fig. 4.4c) and elongation of voids (Fig. 4.4b). Cross or reticulate striation of the plasmic fabric is generally observed in shrink-swell soils (soils with vertic character) with abundant smectite and a coefficient of linear extensibility (COLE) ≥ 0.06 (Soil Survey Staff 2014). The COLE values in the soil control section of these soils are <0.06, which discounts the possibility of stress by shrink-swell. In view of extensive seismic activity (>5 on Richter's scale) in the IGP in the past and during the Holocene (Kumar et al. 1996), these micromorphological features were ascribed to the result of tectonic activity during the

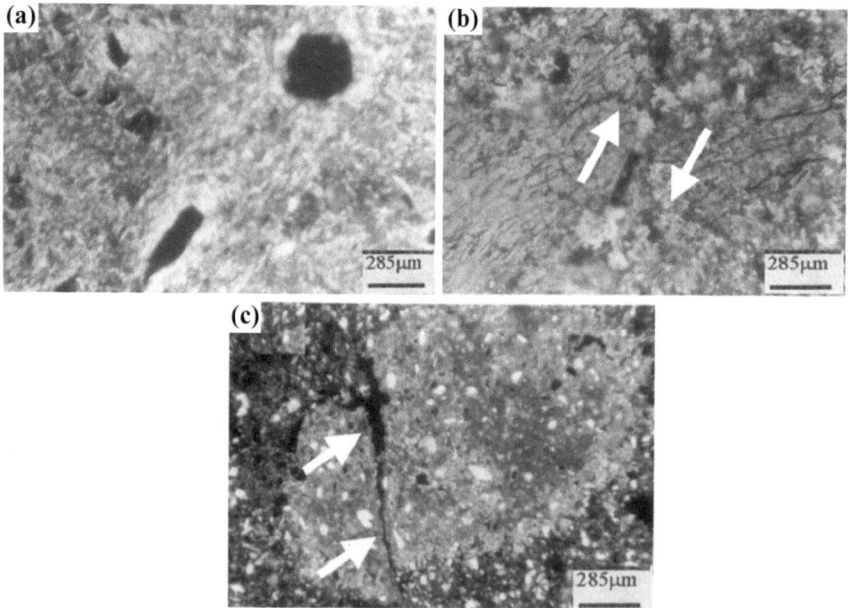

Fig. 4.4 Representative photomicrographs in cross-polarized light of micromorphological features of Haplustalfs and Natrustalfs of the IGP. **a** Moderate to strong cross-striated plasmic fabric. **b** Fragmented and displaced clay pedofeatures and formation of elongated voids. **c** Fragmented and displaced nodules of calcium carbonates (Adapted from Pal et al. 2009b)

Holocene (Pal et al. 2003b). It is worth mentioning here that the striation of plasmic fabric in the form of poro/parallel/unistrial/reticulate, disruption of clay pedofeatures and the presence of irregular, elongated, zigzag and broken voids are common in IGP soils with non-vertic character with COLE < 0.06 (Pal et al. 2010) (Ghabdan, in unit 5; Zarifa Viran, unit16; Simri, unit 32; Belsar, unit 50; Fig. 4.1) as well as in soils with vertic character having COLE > 0.06 (Pal et al. 2010) (Dhadde and Jagjitpur, unit 5; Sakit, unit 23; Konarpara, unit 56; Fig. 4.1), which discounts the exclusive role of shrink-swell to relate the stress. In view of this enigmatic situation a fresh initiative by the soil micro morphologists is warranted so that a selective procedure is evolved as a tool to identify the stress specific to nonpedogenic and pedogenic origin for soils with vertic character.

4.7 Soils of the IGP: Bio-climates, Pedogenic Processes, Mineralogy and Taxonomic Classes

4.7.1 Development of Soils in Different Bio-climates

Soils of the IGP occur in seven bio-climatic zones varying from hot-arid in the west to per humid conditions in the east (Fig. 4.5). The soils in the western part of the IGP with hot-arid climate have sandy loam texture, low organic carbon (OC), are calcareous and moderately alkaline in nature but not sodic due to less leaching (Pal et al. 2010). Further east, the semi-arid dry (SAD) soils are loam to sandy clay in texture, have low OC, and are calcareous and neutral to moderate alkaline in nature (Pal et al. 2010). Some of the SAD soils are also sodic in nature (Pal et al. 2010). In the central and eastern part of the IGP with sub humid dry (SHD) and sub-humid moist (SHM) climate, the soils are silty loam to silty clay loam in texture,

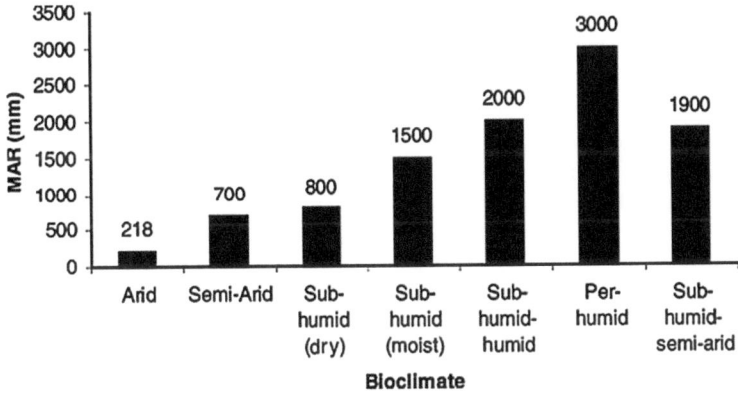

Fig. 4.5 Different bio-climate system of the IGP and distribution of rainfall in each bio-climate (Adapted from Pal et al. 2009b)

impoverished in OC, and moderate alkaline, and less calcareous in nature (Pal et al. 2010). In the southernmost part of the deltaic plains with humid moist (HM) climate, the soils are silty clay to silty clay loam in texture, enriched with OC, and saline-alkali in nature (Pal et al. 2010). Soils over the Tista-Brahmaputra Plains (TBP) are characterized by per humid climate (PH), silty clay loam to sandy clay loam texture, high OC, slight acidity, and no PC (Pal et al. 2010). The prevailing climatic variability across the entire IGP plays a key role in the contemporary pedogenesis and also influenced soils with polygenetic features on older geomorphic surfaces (Srivastava and Parkash 2002; Pal et al. 2010; Srivastava et al. 2015, 2016).

4.7.2 Pedogenic Processes, Polygenesis and Taxonomic Classes of the IGP Soils

Addition and depletion of organic carbon (OC), formation of PC, illuviation of clay particles, and argilli-pedoturbation are the major pedogenic processes in soils of the IGP during the Holocene (Pal et al. 2000a, b, 2003a; Srivastava 2001; Srivastava et al. 2016). These pedogenic processes led to the formation of Entisols, Inceptisols, Alfisols and Vertisols with varying climate and parent material across the IGP. The PC occurring as powdery to large nodules, is marked by a general decrease in subsoils from hot arid to sub humid climates (Pal et al. 2000a; Srivastava 2001) and is even absent in soils of per humid regions (e.g. Seoraguri soils, unit 59, Fig. 4.1). In western part of the IGP, the Entisols and Inceptisols formed on QIG1 and QIG2 surfaces with large amount of PC due to aridity. The Alfisols commonly occur in all climatic zones and correspond to QIG4 and QIG5 geomorphic surfaces of the IGP. In hot-arid and semiarid dry climates such soils are highly alkaline and mostly sodic in nature (Haplargidic/Typic/Vertic Natrustalfs), whereas in SAM, SHD, SAM and PH climates, the soils become mildly alkaline, neutral-acidic, and non-sodic in nature (Oxyaquic/Typic Haplustalfs, Aeric/Vertic/Typic/Umbric Endoaqualfs) (Pal et al. 2010). The increased flux of Cratonic origin enriched with plagioclase and smectitic clay as its weathering product has resulted vertic features in many soils in Bihar (Ekchari, Vertic Endoaqualfs and Sarthua, Vertic Endoaqualfs) and in West Bengal at Chunchura (unit 57, Fig. 4.1; Typic Endoaquerts), Hangram (unit 56, Fig. 4.1;Vertic Endoaqualfs), Konarpara (unit 56, Fig. 4.1; Vertic Endoaqualfs), Madhpur (unit 56, Fig. 4.1; Vertic Endoaqualfs), Sasanga(unit 56, Fig. 4.1; Chromic Vertic Endoaqualfs), Mohanpur (unit 56, Fig. 4.1; Vertic Endoaqualfs) and Sagar (unit 56, Fig. 4.1; Vertic Endoaqualfs) (Pal et al. 2010). Soils with vertic character in Uttar Pradesh (Hirapur, Vertic Natrustalfs and Sakit, Oxyaquic Vertic Haplustalfs) (unit 23, Fig. 4.1) and Punjab (Dhadde of unit 5, Fig. 4.1; Oxyaquic Vertic Haplustalfs and Jagjitpur of unit 5, Fig. 4.1; Oxyaquic Vertic Haplustalfs) have also received smectitic mineral as alluvium brought through rivers originated at base rich rocks (Ray et al. 2006; Pal et al. 2010).

(a)   (b)

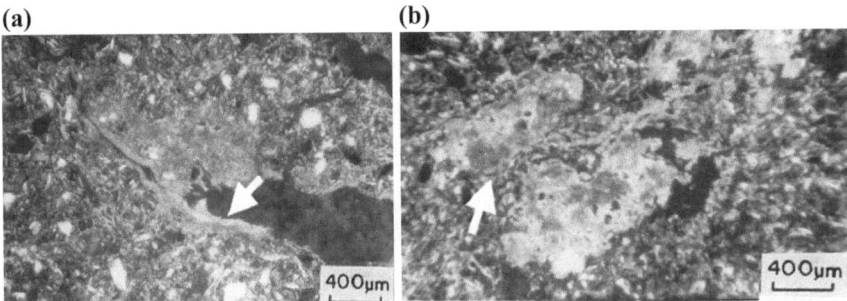

Fig. 4.6 Representative photomicrographs in cross-polarized light of **a** clay pedofeatures with micrite hypo-coatings and **b** clay pedofeatures between two calcium carbonate nodules in Alfisols of the IGP (Adapted from Pal et al. 2009b)

Most of the IGP soils, except than those with vertic characters, are marked by weathering of biotite on a large scale (Pal et al. 1994, 2003a; Srivastava et al. 1998; Srivastava and Parkash 2002). Instead of pure void argillans, impure clay pedofeatures (please refer to Fig. 3.1b) with impairment of parallel orientation of clay platelets along the voids are typical of these soils (Pal et al. 1994, 2003a, b, 2009a, b, 2010). The presence of clay pedofeatures in soils with vertic characters implies that clay illuviation is an important pedogenic process even in soils with shrink-swell phenomena (Pal et al. 2009a, 2012a; Srivastava et al. 2016). Presence of illuvial clay features together with PC (Fig. 4.6a) in soils indicates that illuviation of clay particles, especially in dry climates occurred under a favourable pH condition, which was higher than the zero point of charge required for complete dispersion of clay caused by the precipitation of soluble Ca^{2+} ions as $CaCO_3$ (Eswaran and Sys 1979). This discounts any role of soluble Ca^{2+} ions and the presence of $CaCO_3$ in preventing the illuviation and the accumulation of clay particles. Thus the formation of illuvial clay pedofeatures and the pedogenic carbonates are two pedogenetic processes that occurred simultaneously during the formation of the IGP soils. The two contemporary pedogenetic events that cause concomitant development of soil sodicity exemplify the pedogenic threshold in dry climates during the Late Holocene (Pal et al. 2003b, 2009a, b, 2012a).

Illuviation of clay particles usually results in the development of clay skins that can be recognized in the field with a 10× lens. However, clay skins cannot be seen clearly in many of the pedons despite >30% increase in total clay in the Bt horizon. Difficulty in identifying clay skins in the clay-enriched B horizons is common in the IGP soils (Pal et al. 1994) and elsewhere (McKeague et al. 1981) under semi-arid climate. However, thin sections of these soils do show the presence of impure clay pedo-features that provides incontrovertible evidence of clay illuviation in sodic environment. It then follows that illuviation of fine clay under ustic moisture regime and mainly in loamy-textured parent material has not always resulted in the presence of clay skins or, where present, in pure void argillans (Pal et al. 1994, 2003a, b). Employing thin section studies to confirm the clay illuviation process is not an easy

Fig. 4.7 Schematic depth distribution of clay mica and vermiculite plus smectite in Alfisols of the IGP. *M* mica; *Vm* vermiculite; *Sm* smectite; *ESP* exchangeable sodium percentage (Adapted from Pal et al. 2003b)

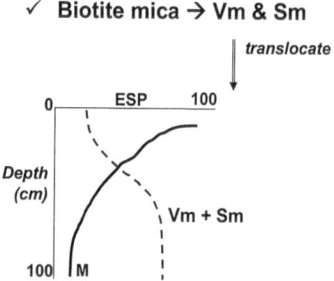

proposition for Indian soil survey programmes. In absence of such analytical tool's facility to confirm the clay illuviation process in soils with clay enriched B horizons are often placed in Inceptisol order, which undermines the contemporary pedogenic process on a stable geomorphic surface. Interestingly, such soils indicate a decrease in clay mica (<2 μm) and an increase in vermiculite and smectite (as the weathered products of biotite mica during the soil formation) with pedon depth (Fig. 4.7), determined both semi-quantitatively (Pal et al. 1994, 2003b) and quantitatively (Pal 1997), which can be a sure test of clay illuviation even when clay skins identifiable in the field are absent.

The Holocene climatic changes caused polypedogenesis in soils that formed on QIG4 and QIG5 surfaces of the IGP (Srivastava and Parkash 2002). Such soils over the western part with arid and semi-arid climatic zones are marked by several episodes of pedogenic carbonates features together with illuvial features (Fig. 4.6b; Srivastava and Parkash 2002; Srivastava et al. 2016). By contrast, the polypedogenic features in soils of the sub-humid to humid central and eastern parts of the IGP are dominated by the illuvial features over the pedogenic carbonates of different phases (Srivastava and Parkash 2002; Srivastava et al. 2016).

Notable changes in characteristics of the IGP soils are noticed after the reclamation and amelioration practices. For example, the Sakit soils which were sodic (Murthy et al. 1982), following reclamation with gypsum and rice as first crop the soils now qualify as Haplustalfs (Pal et al. 2010)(unit 23, Fig. 4.1). However, two soils, namely, Zarifa Viran and Hirapur still remain as sodic soils (Natrustalfs) (Pal et al. 2010). Similarly, Haldi soils which were Mollisols (Murthy et al. 1982); after about two-and-half decades of agricultural practices qualify as Typic Haplustalfs (Pal et al. 2010) (unit 27, Fig. 4.1).

4.7.3 Clay Minerals and Mineralogy Class of the IGP Soils

Majority of the IGP soils contain fine-grained micas (Fig. 4.8) and soil clays are micaceous (e.g. Fig. 4.9; Zarifa Viran Pedon, unit 16), whereas Vertisols and the soils with the vertic character are smectitic (e.g. Fig. 4.10, Chunchura Pedon, unit 57). It is observed that smectitic Vertisols and its vertic intergrades are not developed

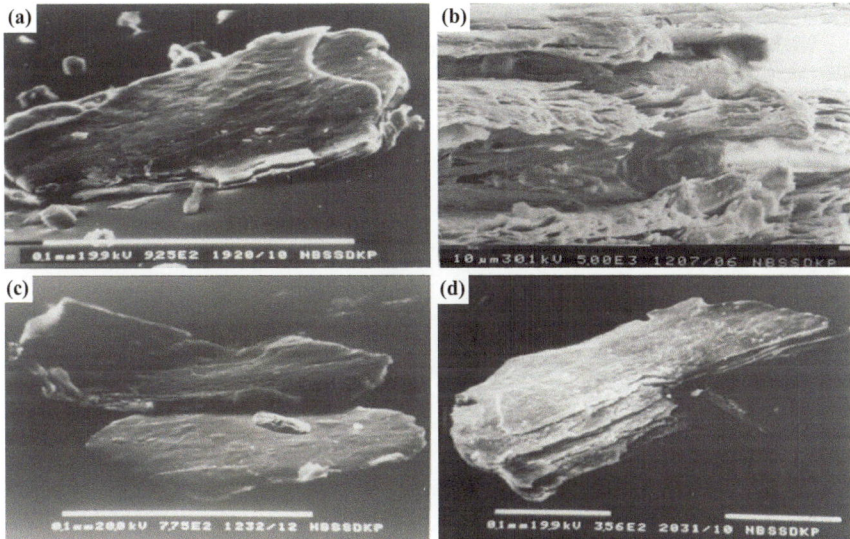

Fig. 4.8 Representative scanning electron microscopic photographs of fine-grained biotites in the semi-arid (**a, b**) and sub-humid (**c, d**) parts of the IGP (Adapted from Pal et al. 2009b)

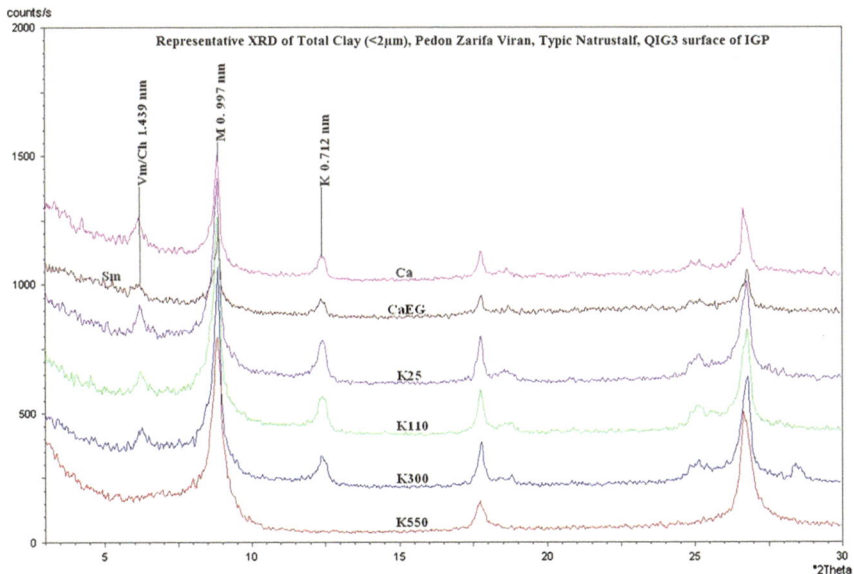

Fig. 4.9 Representative XRD patterns of the total clay fractions of a benchmark soils from semi-arid climate(Zarifa Viran pedon, Typic Natrustalfs) *Sm* smectite (HCS), *V* vermiculite (HIV), *Ch* chlorite (PCh), *M* mica, *K* kaolin Ca-EG = Ca-saturated and glycolated; K25/110/300/550 °C = K-saturated and heated (Adapted from Pal et al. 2010)

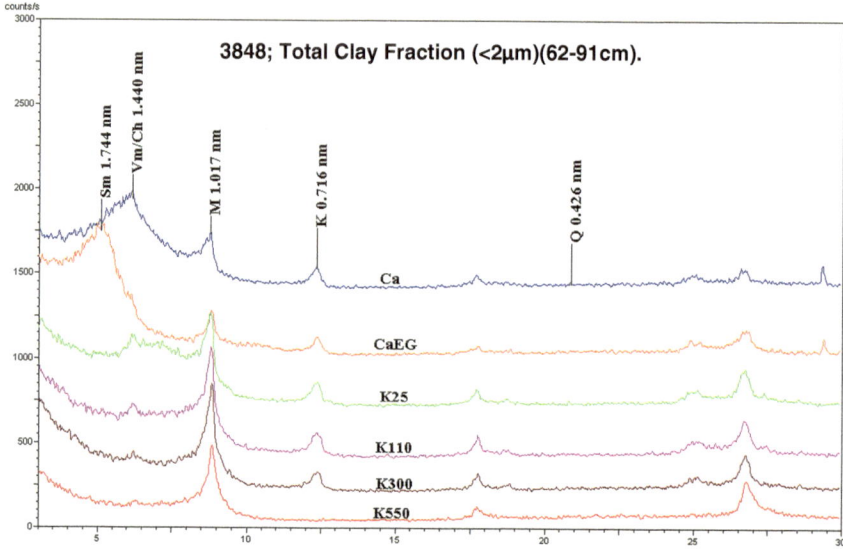

Fig. 4.10 Representative XRD pattern of the total clay fractions of Vertisols of Chunchura. *Sm* smectite (LCS), *V* vermiculite (HIV), *Ch* chlorite (PCh), *M* mica, *K* kaolin, *Q* quartz, *Ca* Ca-saturated; Ca-EG = Ca-saturated and glycolated; K25/110/300/550 °C = K-saturated and heated to 25, 110,300 and 550 °C (Adapted from Pal et al. 2010)

from micaceous alluvium material. If they would have parental legacy to micaceous minerals, all other non-vertic soils should also have smectitic mineralogy instead of being micaceous. The formation of the vertic characters in the soils of the IGP can be explained only in view of smectitic sedimentary flux of Cratonic origin (Ray et al. 2006; Pal et al. 2012a, b). Detailed micromorphological and mineralogical investigations indicate that dominance of the Cratonic flux enriched with plagioclase and smectitic clay has caused the vertic characters in these soils (Singh et al. 1998; Pal et al. 2010, 2012b; Srivastava et al. 2010, 2015, 2016). Further studies indicate that the alluvium of the Himalayan origin contains considerably less smectite (low charge dioctahedral smectite, LCS) as compared to that of the Craton (Pal et al. 2012b). These soils, therefore, contain less amount of fine clay LCS (15–20%) as compared to 25–36% fine clay LCS in vertic intergrades and Vertisols (Pal et al. 2010). They are formed in the alluvium brought down from the Himalayan hinterland that contains <2% plagioclase (Srivastava and Parkash 2002; Srivastava et al. 2010; Pal et al. 2012b). Therefore, the LCS in the IGP soils is the alteration product of plagioclase during earlier humid climate in the source area as the formation of LCS from mica is improbable in humid climate (Pal et al. 2012a, b). Micaceous IGP soils of primarily of the Himalayan source also contain vermiculite, hydroxy-interlayered vermiculite (HIV), pseudo-chlorite (PCh) and kaolin in both silt and clay fractions, and in the clay fractions presence of trioctahedral smectite, HIV, PCh and smectite-kaolin interstratified minerals (Sm/K) is common (Pal et al. 2000b).

Genesis and transformation of many of these minerals (trioctahedral smectite, HIV, PCh and Sm/K) at the expense of biotite mica do not represent the contemporary pedogenesis in slight to moderate alkaline chemical environment induced by semi-arid climate but are often used as evidence of climate change in the geological past (Pal et al. 1989; Srivastava et al. 2013; Pal 2014), details of which are provided in Chap. 7.

References

Bhattacharya T, Banerjee SN (1979) Quaternary geology and geomorphology of the Ajay-Bhagirathi Valley, Birbhum and Murshidabad districts. Indian J Earth Sci 6:91–102

Bhattacharyya T, Pal DK, Chandran P, Mandal C, Ray SK, Gupta RK, Gajbhiye KS (2004) Managing soil carbon stocks in the Indo-Gangetic Plains, India. Rice-wheat consortium for the Indo-Gangetic Plains, New Delhi, p 44. (http://www.rwc-prism, www.cgiar.org and http://www.cimmyt.org)

Birkeland PW (1990) Soil development on stable landforms and implications for landscape studies. Geomorphology 3:207–224

Chen S, Lin B, Baig M, Mitra B, Lopes RJ, Santos AM, Magee DA, Azevedo M, Tarroso P, Sasazaki S, Ostrowski S, Mahgoub O, Chaudhuri TK, Zhang Y, Costa V, Royo LJ, Goyache F, Luikart G, Boivin N, Fuller DQ, Mannen H, Bradley DG, Beja-Pereira A (2010) Zebu cattle are an exclusive legacy of the South Asia Neolithic. Mol Biol Evol 27:1–6

Eswaran H, Sys C (1979) Argillic horizon in LAC soils formation and significance to classification. Pédologie 29:175–190

Fuller DQ (2006) Agricultural origins and frontiers in South Asia: A working synthesis. J World Prehistory 20:1–86

Gaina C, Muller RD, Brown B, Ishihara T (2007) Breakup and early seafloor spreading between India and Antarctica. Geophys J Int 170:151–169

Gaur VK (1994) Evaluation of seismic hazard in India towards minimizing earthquake risk. Curr Sci 67:324–329

Gibling MR, Sinha R, Roy NG, Tandon SK, Jain M (2008) Quaternaryfluvial and eolian deposits on the Belan River, India: paleoclimatic setting of Paleolithic to Neolithic archaeological sites over the past 85,000 years. Quat Sci Rev 27:392–411

Gupta HP (1978) Holocene Palynology from Meander Lake in the Ganga Valley, district Partapgarh, Uttar Pradesh. Palaeobotonist 25:109–119

Harden J (1990) Soil development on stale landforms and implications for landscape studies. Geomorphology 3:391–397

Huggett RJ (1998) Soil chronosequences, soil development, and soil evolution: a critical review. Catena 32:155–172

Kent DV, Muttoni G (2008) Equatorial convergence of India and early Cenozoic climate trends. Proc Natl Acad Sci 105:16065–16070

Khan MSH, Parkash B, Kumar S (2005) Soil-landform development of a part of the fold belt along the eastern coast of Bangladesh. Geomorphology 71:310–327

Klootwijk CT, Gee JS, Peirce JW, Smith GM, McFadden PI (1992) An early India-Asia contact: paleomagnetic constraints from ninety east ridge, ODP leg 121. Geology 20:395–398

Kumar S, Parkash B, Manchanda ML, Singhvi AK, Srivastava P (1996) Holocene landform and soil evolution of the western Gangetic Plains: implications of neotectonics and climate. Zeitschrift fuer Geomorphologie (Suppl.) 103:283–312

Kumar P, Yuan X, Kumar MR, Kind R, Li X, Chadha RK (2007) The rapid drift of the Indian Tectonic Plate. Nature 449:894–897

Leather JW (1898) On the composition of Indian soils. Agric Ledger 4:81–164

Mandal C, Mandal DK, Bhattacharyya T, Sarkar D, Pal DK et al (2014) Revisiting agro-ecological sub-regions of India—a case study of two major food production zones. Curr Sci 107: 1519–1536

McFadden LD, Knuepfer PLK (1990) Soil geomorphology: the linkage of pedology and surficial processes. Geomorphology 3:197–205

McKeague JA, Wang C, Ross GJ, Acton CJ, Smith RE, Anderson DW, Petapiece WW, Lord TM (1981) Evaluation of criteria for argillic horizons (Bt) of soils in Canada. Geoderma 25:63–74

McNeill JR, Winiwarter V (2004) Breaking the SOD: humankind, history and soil. Science 304:1627–1629

Meadow R, Patel AK (2003) Prehistoric pastoralism in North-western South Asia from the Neolithic through the Harappan period. In: Weber SA, Belcher WR (eds) Indus ethnobiology. Lexington Books, Lanham, New perspectives from the field, pp 65–94

Misra VN (2001) Prehistoric human colonization of India. J Biosci 26:491–531

Mohindra R, Parkash B, Prasad J (1992) Historical geomorphology and pedology of the Gandak Megafan, Middle Gangetic Plains, India. Earth Surf Proc Land 17:643–662

Murthy RS, Hirekerur LR, Deshpande SB, Venkat Rao BV (eds) (1982) Benchmark soils of India. National Bureau of Soil Survey and Land Use Planning, Nagpur, India (374 pp)

Pal DK (1997) An improvised method to identify clay illuviation in soils of Indo-Gangetic plains. Clay Res 16:46–50

Pal DK (2014) Minerals in soils and sediments as evidence of climate change: a review. Gondwana Geol Mag 29:87–94

Pal DK, Deshpande SB, Venugopal KR, Kalbande AR (1989) Formation of di- and trioctahedral smectite as evidence for paleo-climatic changes in southern and central peninsular India. Geoderma 45:175–184

Pal DK, Kalbande AR, Deshpande SB, Sehgal JL (1994) Evidence of clay illuviation in sodic soils of north-western part of the Indo-Gangetic Plains since the Holocene. Soil Sci 158:465–473

Pal DK, Bhattacharyya T, Deshpande SB, Sarma VAK, Velayutham M (2000a) Significance of minerals in soil environment of India, NBSS review series 1. NBSS & LUP, Nagpur

Pal DK, Dasog GS, Vadivelu S, Ahuja RL, Bhattacharyya T (2000b) Secondary calcium carbonate in soils of arid and semi-arid regions of India. In: Lal R, Kimble JM, Eswaran H, Stewart BA (eds) Global climate change and pedogenic carbonates. Lewis Publishers, Boca Raton, Fl, pp 149–185

Pal DK, Srivastava P, Durge S, Bhattacharyya T (2003a) Role of micro topography in the formation of sodic soils in the semi-arid part of the Indo-Gangetic Plains, India. Catena 51:3–31

Pal DK, Srivastava P, Bhattacharyya T (2003b) Clay illuviation in calcareous soils of the semi-arid part of the Indo-Gangetic Plains, India. Geoderma 115:177–192

Pal DK, Bhattacharyya T, Chandran P, Ray SK (2009a) Tectonics-climate-linked natural soil degradation and its impact in rainfed agriculture: Indian experience. In: Wani SP, Rockström J, Oweis T (eds) Rainfed agriculture: unlocking the potential. CABI International, Oxfordshire, U.K., pp 54–72

Pal DK, Bhattacharyya T, Srivastava P, Chandran P, Ray SK (2009b) Soils of the Indo-Gangetic Plains: their historical perspective and management. Curr Sci 9:1193–1201

Pal DK, Lal S, Bhattacharyya T, Chandran P, Ray SK, Satyavathi PLA, Raja P, Maurya UK, Durge SL, Kamble GK (2010) Pedogenic thresholds in benchmark soils under rice-wheat cropping system in a climosequence of the Indo-Gangetic Alluvial Plains. Final Project Report, Division of Soil Resource Studies. NBSS & LUP, Nagpur (193 pp)

Pal DK, Bhattacharyya T, Sinha R, Srivastava P, Dasgupta AS, Chandran P, Ray SK, Nimje A (2012a) Clay minerals record from Late Quaternary drill cores of the Ganga Plains and their implications for provenance and climate change in the Himalayan Foreland. Palaeogeogr Palaeoclimatol Palaeoecol 356–357:27–37

Pal DK, Wani SP, Sahrawat KL (2012b) Vertisols of tropical Indian environments: pedology and edaphology. Geoderma 189–190:28–49

Parkash B, Kumar S (1991) Indo-Gangetic Basin. In: Tandon SK, Pant CC, Kashyap SM (eds) Sedimentary basins of India, tectonic context. Gyanodaya Prakashan, Nainital, pp 147–170

Parkash B, Kumar S, Rao MS, Giri SC, Kumar CS, Gupta S, Srivastava P (2000) Holocene tectonic movements and stressfield in the Western Gangetic Plains. Curr Sci 79:438–449

Patel AK (1999) New radiocarbon determinations from Loteshwar and their implications for understanding settlement and subsistence in North Gujarat and adjoining areas. Paper presented in the fifteenth international conference on South Asian archaeology, Leiden University, July 5–9, p 1999

Raiverman V, Kunte S, Mukherjee A (1983) Basin geometry, Cenozoic sedimentation and hydrocarbon prospects in northwestern Himalaya and Indo-Gangetic Plains. Pet Asia J 6:67–92

Rao MBR (1973) The subsurface geology of Indo-Gangetic Plains. J Geol Soc India 14:217–242

Ray SK, Bhattacharyya T, Chandran P, Sahoo AK, Sarkar D, Durge SL, Raja P, Maurya UK, Pal DK (2006) On the formation of cracking clay soils (Vertisols) in West Bengal. Clay Res 25:141–152

Ritter DF(1986) Process geomorphology. Wm C Brown, Dubuque Iowa (603 pp)

Ruhe RV (1969) Quaternary landscapes in Iowa. Iowa State University Press, Ames (255 pp)

Sastri VV, Venkatachala BS, Bhandari LL, Raju ATR, Datta AK (1971) Tectonic framework and subsurface stratigraphy of the Ganga Basin. J Geol Soc India 12:222–233

Saxena A, Prasad V, Singh IB, Chauhan MS, Hasan R (2006) On the Holocene record of phytoliths of wild and cultivated rice from Ganga Plain: evidence for rice-based agriculture. Curr Sci 90:1547–1552

Shankarnarayana HS, Sarma VAK (1982) Soils of India. In: Murthy RS et al (eds) Benchmarks soils of India–morphology, characteristics and classification for resource management. NBSS & LUP, Nagpur, pp 41–69

Sharma GR, Misra VD, Mandal D, Misra BB, Pal JN (1980) Beginning of agriculture. Avinash Prakashan, Allahabad, pp 133–200

Singh IB (1988) Geological evolution of Gangetic Plains-What we know and what we do not know?. Lucknow University, Lucknow (Abstract volume), Workshop on Gangetic Plains-Terra Incognita

Singh LP, Parkash B, Singhvi AK (1998) Evolution of the lower Gangetic Plain landforms and soils in West Bengal, India. Catena 33:75–104

Singhai SK, Parkash B, Manchanda M (1991) Geomorphological and pedological evolution of Haryana state. Bull Oil Nat Gas Comm 28:37–60

Soil Survey Staff (2014) Keys to soil taxonomy, twelfth edn. United States Dept. of Agriculture, Natural Resource Conservation Service, Washington, DC

Srivastava P (2001) Paleoclimatic implications of pedogenic carbonate in Holocene soils of the Gangetic Plains. Palaeogeogr Palaeoclimatol Palaeoecol 172:207–259

Srivastava P, Parkash B (2002) Polygenetic soils of the north-central part of the Gangetic Plains: a micromorphological approach. Catena 46:243–259

Srivastava P, Parkash B, Sehgal JL, Kumar S (1994) Role of neotectonics and climate in development of the Holocene geomorphology and soils of the Gangetic Plains between Ramganga and Rapti rivers. Sed Geol 94:119–151

Srivastava P, Parkash B, Pal DK (1998) Clay minerals in soils as evidence of Holocene climatic change, central Indo-Gangetic Plains, north-central India. Quatern Res 50:230–239

Srivastava P, Rajak MK, Sinha R, Pal DK, Bhattacharyya T (2010) A high resolution micromorphological record of the late quaternary Paleosols from Ganga-Yamuna interfluve: stratigraphic and paleoclimatic implications. Quatern Int 227:127–142

Srivastava P, Pal DK, Bhattacharyya T (2013) Mineral formation in soils and sediments as signatures of climate change. In: Bhattacharyya T, Pal DK, Sarkar D, Wani SP (eds) Climate change and agriculture. Studium Press, New Delhi, pp 223–234

Srivastava P, Pal DK, Aruche KM, Wani SP, Sahrawat KL (2015) Soils of the Indo-Gangetic Plains: a pedogenic response to landscape stability, climatic variability and anthropogenic activity during the Holocene. Earth Sci Rev 140:54–71. doi:10.1016/j.earscirev.2014.10.010

Srivastava P, Aruche M, Arya A, Pal DK, Singh LP (2016) A micromorphological record of contemporary and relict pedogenic processes in soils of the Indo-Gangetic Plains: implications for mineral weathering, provenance and climatic changes. Earth Surf Proc Land 41:771–790. doi:10.1002/esp.3862

Velayutham M, Pal DK (2004) Soil classification, India. Encyclopedia of Soil Science. Marcel Dekker, New York (1–3 pp)

Velayutham M, Pal DK, Bhattacharyya T, Srivastava P (2002) Soils of the Indo-Gangetic Plains, India–the historical perspective. In: Abrol YP, Sangwan S, Tiwari M (eds) Land use–historical perspectives–Focus on Indo-Gangetic Plains. Allied Publishers, New Delhi, pp 61–70

Voelcker JA (1893) Improvement of Indian agriculture. Report submitted to Famine Commission of 1880. Imperial and Provincial Agricultural Department, India

Vreeken WJ (1975) Principal kind of soil chronosequences and their significance in soil history. J Soil Sci 26:378–390

Wadia DN (1966) Geology of India, 6th ed. McMillan, London (531 pp)

Wadia DN, Krishnan MS, Mukherjee PN (1935) Introductory note on the geological formation of the soils of India. Record Geol Surv India 68:369–391

Wasson RJ (2006) Human interaction with soil-sediment systems in Australia. In: McNeil JR, Winiwarter V (eds) Towards a world environmental history of soils. Oregon State University Press, Cambridge, pp 243–272

Williams MAJ, Clarke MF (1984) Late Quaternary environments in North-Central India. Nature 308:633–635

Williams MAJ, Clarke MF (1995) Quaternary geology and prehistoric environments in the Son and Belan valleys, north central India. Memoir Geol Surv India 32:282–308

Williams MAJ, Pal JN, Jaiswal M, Singhvi AK (2006) River response to Quaternary climatic fluctuations: evidence from the Son and Belan valleys, north-central India. Quatern Sci Rev 25:2619–2631

Chapter 5
Conceptual Models on Tropical Soil Formation

Abstract Use of models to explain adequately the formation of tropical soils indicates that although among the most popular models applicable in soil formation, the residua and haplosoil models have relevance to formation and persistence of Indian tropical soils, they cannot explain the existence of million years old Vertisols, Alfisols and Mollisols under humid tropical climate because these models did not consider the stability of base rich primary minerals over time. This novel understanding provides a deductive check on the inductive reasoning so far made on the formation of soils in tropical humid climate and also establishes the validity of Jenny's state factor equation in the formation of the Indian tropical soils in the intense weathering environments under HT climate.

Keywords Validation of models · Acidic Alfisols · Acidic Vertisols · Acidic Mollisols · Ultisols

5.1 Introduction

Dijkerman (1974) listed four types of conceptual models. They are (1) mental, (2) verbal, (3) structural and (4) mathematical. Whereas most of the working models are of a verbal nature, more emphasis is now being laid in mathematical models due to availability of computers (Smeck et al. 1983). It would be worthwhile exercise to examine the pros and cons of the most popular model employed in pedogenetic studies of Indian tropical soils (spatially associated Mollisols-Alfisols-Vertisols) in order to gain a better understanding of factors and processes operative in tropical soil system. Among the models known to us are (1) state-factor analysis (Jenny 1941), (2) energy model-factorial model (Runge 1973), (3) residua and haplosoil models (Chesworth 1973a, 1980), (4) generalized process model (Simonson 1959) and (5) soil-landscape model (Huggett 1975). Out of these models the most challenging ones are residua and haplosoil models (Chesworth 1973a, 1980) and have strong relevance to the formation of Vertisols, Alfisols, Mollisols and Ultisols in humid tropical climate of India. Formation and persistence of such

© Springer International Publishing AG 2017
D.K. Pal, *A Treatise of Indian and Tropical Soils*,
DOI 10.1007/978-3-319-49439-5_5

soil orders under prolonged HT climate provides an opportunity to validate both residua and haplosoil system, and in this exercise soils developed on zeolitic Deccan basalts, gneissic and sedimentary rock systems are discussed in the following.

5.2 Vertisols Spatially Associated with Alfisols and Mollisols on Zeolitic Deccan Basalts

Vertisols, Alfisols, and Mollisols are the members of Mollisol-Alfisol-Vertisol association (Bhattacharyya et al. 2005, 2006) on the zeolitic Deccan basalt areas. The associated Alfisols were formed in HT climate and are persisting since the early Tertiary (Bhattacharyya et al. 1999). The transformation of smectite (the first weathering product of the Deccan basalt) (Pal and Deshpande 1987) to kaolin (Sm-K) during HT weathering began at the end of the Cretaceous and continued during the Tertiary (Kumar 1986), and thus Alfisols date back to the Tertiary and the Cretaceous (Idnurm and Schmidt 1986). With a combination of high temperature and adequate moisture, the HT climate of the Western Ghats and Satpura Range of central India provided a weathering environment that should have nullified the effect of parent rock composition in millions of years, resulting in kaolinitic and/or oxidic mineral assemblages consistent with either residua (Chesworth 1973a) or haplosoil (Chesworth 1980) models of tropical soil formation like in Ultisols and Oxisols (Soil Survey Staff 1999). Instead, the soils of the zeolitic Deccan basalt have Sm-K and represent Vertisols, Alfisols and Mollisols. The knowledge gained on the role of zeolites in the persistence of soils not only provides a deductive check on the inductive reasoning on the formation of soil in the HT climate, but also throws light on the role of these minerals in preventing loss of soil productivity even in an intense leaching environment. This indeed may be the reason why crops do not show response to liming in acid soils of the tropical Western Ghats (Kadrekar 1979).

5.3 Acidic Alfisols, Mollisols and Ultisols on Gneissic and Sedimentary Rock Systems

Under Indian HT climate, soils formed do not belong to the same soil order; they belong to Ultisols, Alfisols and Mollisols. Mollisols are prevalent under thick forest vegetation, whereas Alfisols are under sparse forest vegetation and/or in agricultural soils and Ultisols are in general under agriculture (Pal et al. 2014). However, there are strongly acidic Ultisols and mildly acidic to neutral Alfisols and Mollisols.

In Goa on gneissic rock, strongly acidic Ultisols are associated with moderately acidic Alfisols (Chandran et al. 2004; Harindranath et al. 1999). In Karnataka, on gneissic rock in the Western Ghats area, moderately acidic Ultisols, Alfisols and Mollisols are spatially associated (Shiva Prasad et al. 1998). In the Nilgiri Hills areas of Tamil Nadu on gneissic rock, strongly acidic Ultisols are spatially associated with mildly acidic Alfisols and Mollisols (Natarajan et al. 1997). In Nilgiri hills areas in Kerala on calc-gneiss, near the north of the Palghat Gap strongly acidic Ultisols are associated with mildly acidic to neutral Mollisols, whereas one of the major soil orders in the Palghat Gap is mildly acidic Alfisols (Krishnan et al. 1996).

In NEH areas on sedimentary and gneissic rock, strongly acidic Ultisols are associated with moderately acidic Alfisols (Bhattacharyya et al. 2000; Maji et al. 2000, 2001; Nayak et al. 1996b; Sen et al. 1996, 1999; Singh et al. 1999). In the Andaman and Nicobar Islands on calcareous/micaceous sandstones and lime stones, neutral to slightly alkaline Mollisols are associated with slightly acidic pH Alfisols (Das et al. 1996).

5.4   Critique on the Validation of the Conceptual Models

The soils of the zeolitic Deccan basalt, gneiss/calc-gneiss, sedimentary deposits, lime stones have kaolin and other hydroxy-interlayered clay minerals in abundance (details are available in Chaps. 2 and 3) and represent Ultisols, Alfisols and Mollisols, suggesting that the presence of Ca-bearing weatherable minerals in the soil parent materials under forest vegetation have influenced the weathering rate and exerted a decisive influence on the nature of the soil silicate clay minerals and the formation of soils with various soil orders (Pal et al. 1989). It follows from these results that factors and processes of formation and persistence of these three soil orders in adverse HT climate during the Cenozoic time (Pal et al. 2014) need a further analysis in the light of the existing conceptual models for tropical soil formation.

In either residual (Chesworth 1973a) or haplosoil model of soil formation (Chesworth 1980) in tropical climate, it was envisaged that with a combination of high temperature and adequate moisture, the HT climate of India provided a weathering environment that should have nullified the effects of parent material composition by resulting kaolinitic and/or oxidic mineral assemblages (Chesworth 1973b). The models of Chesworth were based on the hypothesis that (a) the effect of parent rock will be overshadowed and nullified with time; (b) its effect will be evident only in younger or relatively immature soils, and (c) the time is the only independent variable of soil formation or any other processes occurring spontaneously in nature.

In soils of HT climate of India, the dominance of kaolin indicates that in spite of prolonged weathering for millions of years since the early Tertiary, the weathered products of Ca-rich minerals of the parent materials have not reached even the pure kaolinitic mineral stage. The formation of kaolin clay mineral suggests that the formation of Vertisols, Ultisols, Alfisols and Mollisols and their pedogenic threshold at this time supports the supposition that steady state may exist in soils developed over long periods of time spanning not just thousands of years (Smeck et al. 1983; Yaalon 1971, 1975) but also millions of years (Bhattacharyya et al. 1993, 1999; Chandran et al. 2005). The hypothesis of Chesworth for soil formation in HT climate of India cannot explain the persistence of Vertisols, Ultisols, Alfisols and Mollisols, because of the stability of feldspar, zeolites and other Ca-rich minerals/rock (i.e. limestone, and calc-gneiss) over time was not considered in his model. Therefore, the formation and persistence of these soils provide an example that in an open system such as the soil, the existence of a steady state seems a more useful concept than based on equilibrium in a rigorous thermodynamic sense (Smeck et al. 1983; Bhattacharyya et al. 1999, 2006; Chandran et al. 2005), and Jenny's state factor equation is also essentially valid. This contention finds support from the current pedogenetic processes in Ultisols in NEH areas. The OC rich Ultisols have less Al-saturation in surface horizons due to the downward movement of Al as organo-metal complexes or chelates, but have higher base saturation than the sub-surface horizons due to addition of alkaline and alkaline metal cations (please refer to Table 3.2) through litter fall (Nayak et al. 1996a), and there is no desilication and transformation of kaolin to gibbsite (Pal et al. 2014).

In view of contemporary pedogenesis, it is difficult to reconcile that Ultisols (especially of Kerala hitherto considered to be of international reference for laterite) would ever be weathered to reach the Oxisols stage with time frame as envisaged by Smeck et al. (1983), Lin (2011) (please refer to Fig. 2.12a). The knowledge gained on the role of Ca-rich parent materials and Ca-zeolites in the persistence of soils for millions of years provide a deductive check on the inductive reasoning on the formation of Vertisols, Alfisols, Mollisols and Ultisols in the HT climate.

References

Bhattacharyya T, Pal DK, Deshpande SB (1993) Genesis and transformation of minerals in the formation of red (Alfisols) and black (Inceptisols and Vertisols) soils on Deccan Basalt in the Western Ghats, India. J Soil Sci 44:159–171

Bhattacharyya T, Pal DK, Srivastava P (1999) Role of zeolites in persistence of high altitude ferruginous Alfisols of the Western Ghats, India. Geoderma 90:263–276

Bhattacharyya T, Pal DK, Srivastava P (2000) Formation of gibbsite in presence of 2:1 minerals: an example from Ultisols of northeast India. Clay Miner 35:827–840

Bhattacharyya T, Pal DK, Chandran P, Ray SK (2005) Land-use, clay mineral type and organic carbon content in two Mollisols-Alfisols-Vertisols catenary sequences of tropical India. Clay Res 24:105–122

Bhattacharyya T, Pal DK, Lal S, Chandran P, Ray SK (2006) Formation and persistence of Mollisols on zeolitic Deccan basalt of humid tropical India. Geoderma 136:609–620

Chandran P, Ray SK, Bhattacharyya T, Dubey PN, Pal DK, Krishnan P (2004) Chemical and mineralogical characteristics of ferruginous soils of Goa. Clay Res 23:51–64

Chandran P, Ray SK, Bhattacharyya T, Srivastava P, Krishnan P, Pal DK (2005) Lateritic soils of Kerala, India: their mineralogy, genesis and taxonomy. Aust J Soil Res 43:839–852

Chesworth W (1973a) The residua system of chemical weathering: a model for the chemical breakdown of silicate rocks at the surface of the earth. J Soil Sci 24:69–81

Chesworth W (1973b) The parent rock effect in the genesis of soil. Geoderma 10:215–225

Chesworth W (1980) The haplosoil system. Am J Sci 280:909–985

Das AL, Goswami A, Thampi CJ, Sarkar D, Sehgal J (1996) Soils of Andaman and Nicobar Islands for optimising land use, NBSS Publ. 61 (Soils of India Series), National Bureau of Soil Survey and Land Use Planning, Nagpur, India, 57 pp + 5 sheets of soil map (1:50,000 scale)

Dijkerman JC (1974) Pedology as a science: the role of data, models and theories in the study of natural soil systems. Geoderma 11:73–93

Harindranath CS, Venugopal KR, Raghu Mohan NG, Sehgal J, Velayutham MV (1999) Soils of Goa for optimising land use. NBSS publ. 74b (Soils of India Series), National Bureau of Soil Survey and Land Use Planning, Nagpur, India, 131 pp + 2 sheets of soil map on 1:500,000 scale

Hugget RI (1975) Soil landscape systems: amodel of soil genesis. Geoderma 13:1–22

Idnurm M, Schmidt PW (1986) Paleo-magnetic dating of weathered profile. Geol Surv India Memoir 120:79–88

Jenny H (1941) Factors of soil formation. McGraw Hill, NewYork, NY. 28 1 pp

Kadrekar SB (1979) Utility of basic slag and liming material in lateritic soils of Konkan. Indian J Agron 25:102–104

Krishnan P, Venugopal KR, Sehgal J (1996) Soil resources of Kerala for land use planning. NBSS Publ. 48b (Soils of India series 10), National Bureau of Soil Survey and Land Use Planning, Nagpur, India, 54 pp + 2 sheets of soil map on 1:500,000 scale

Kumar A (1986) Palaeo-altitudes and the age of Indian laterites. Palaeogeogr Palaeoclimatol Palaeoecol 53:231–237

Lin H (2011) Three principles of soil change and pedogenesis in time and space. Soil Sci Soc Am J 75:2049–2070

Maji AK, Dubey PN, Verma TP, Chamuah GS, Sehgal J, Velayutham M (2000) Soils of Nagaland: their kinds, distribution, characterisation and interpretations for optimising land use. NBSS publ. 67b. NBSS&LUP, Nagpur, (28 pp)

Maji AK, Dubey PN, Sen TK, Verma TP, Marathe RA, Chamuah GS, Velayutham M, Gajbhiye KS (2001) Soils of Mizoram: their kinds, distribution, characterisation and interpretations for optimising land use. NBSS Publ 75b. NBSS&LUP, Nagpur, (28 pp)

Natarajan A, Reddy PSA, Sehgal J, Velayutham M (1997) Soil resources of Tamil Nadu for land use planning. NBSS publ 46b (Soils of India series), National Bureau of Soil Survey and Land Use Planning, Nagpur, India, 88 pp + 4 sheets of soil map on 1:500,000 scale

Nayak DC, Sen TK, Chamuah GS, Sehgal JL (1996a) Nature of soil acidity in some soils of Manipur. J Indian Soc Soil Sci 44:209–214

Nayak DC, Chamuah GS, Maji AK, Sehgal J, Velayutham M (1996b) Soils of Arunachal Pradesh for optimising land use. NBSS Publ. 55b (Soils of India Series), National Bureau of Soil Survey and Land Use Planning, Nagpur, India, 54 pp + one sheet soil map (1:500,000 scale)

Pal DK, Deshpande SB (1987) Characteristics and genesis of minerals in some benchmark Vertisols of India. Pedologie (Ghent) 37:259–275

Pal DK, Deshpande SB, Venugopal KR, Kalbande AR (1989) Formation of di- and trioctahedral smectite as an evidence for paleoclimatic changes in southern and central Peninsular India. Geoderma 45:175–184

Pal DK, Wani SP, Sahrawat KL, Srivastava P (2014) Red ferruginous soils of tropical Indian environments: a review of the pedogenic processes and its implications for edaphology. Catena: 121:260–278. doi:10.1016/j.catena2014.05.023

Runge ECA (1973) Soil development sequences and energy models. Soil Sci 115:183–193

Sen TK, Chamuah GS, Maji AK, Sehgal J (1996) Soils of Manipur for optimising land use. NBSS publ 56b (Soils of India series), National Bureau of Soil Survey and Land Use Planning, Nagpur, India, 52p + 1 sheet map

Sen TK,Chamuah GS, Sehgal J, Velayutham M (1999) Soils of Assam for optimizing land use. NBSS publ 66b (Soils of India series), National Bureau of Soil Survey and Land Use Planning, Nagpur, India, 51 pp + 2 sheets map

Shiva Prasad CR, Reddy PSA, Sehgal J, Velayutham M (1998) Soils of Karnataka for optimising land use. NBSS publ 47b (Soils of India series), National Bureau of Soil Survey and Land Use Planning, Nagpur, India, 111 pp + 4 sheets of soil map on 1:500,000 scale

Simonson RW (1959) Outline of a generalized of soil genesis. Soil Sci Soc Am Proc 23:152–156

Singh RS, Maji AK, Sehgal J, Velayutham M (1999) Soils of Meghalaya for optimizing land use, NBSS publ 52b (Soils of India series). National Bureau of Soil Survey and Land Use Planning, Nagpur, India, p. 29 + 1 sheet soil map (1:500,000 scale)

Smeck NE, Runge ECA, Mackintosh EE (1983) Dynamics and genetic modelling of soil system. In: Wilding LP, Smeck NE, Hall GF (eds) Pedogenesis and soil taxonomy-concepts and interactions. Elsevier, Amsterdam, Developments in Soil Science II-A, pp 51–81

Soil Survey Staff (1999) Soil taxonomy: a basic system of soil classification for making and interpreting soil surveys, USDA-SCS agricultural handbook no 436, 2nd edn. U.S. Govt, Printing Office, Washington, DC

Yaalon DH (1971) Soil forming processes in time and space. In: Yaalon DH (ed) Paleopedology. Israel University Press, Jerusalem, pp 29–39

Yaalon DH (1975) Conceptual models in pedogenesis. Can soil forming functions be solved? Geoderma 14:189–205

Chapter 6
Land and Soil Degradation and Remedial Measures

Abstract Most published research on soil degradation in general emphasizes the role of anthropogenic factors. Even among the natural soil degradation processes the regressive pedogenic processes that lead to the formation of $CaCO_3$ and concomitant development of subsoil sodicity and the adverse effects of palygorskite mineral on the soils of the semi-arid tropics (SAT), have received little global attention as the natural processes of chemical degradation of soils. Pedogenic calcium carbonate, soil sodicity and palygorskite mineral impair the hydraulic properties of the SAT soils, which reduce their crop productivity. This type of unfavourable soil health triggered by the tectonic-climate linked regressive pedogenic processes (formation of pedogenic calcium carbonate and development of sub soil sodicity) needs to be globally considered as the natural soil degradation process. The regressive pedogenic processes that are inherently connected to the development of natural soil degradation, expands the basic knowledge in pedology and thus it may have relevance in soils of other SAT areas of the world. Research efforts made in the Indian subcontinent explains the cause-effect relationship of the degradation and provides enough insights as to how the remedial measures are to be invented including the role of pedogenic $CaCO_3$ and geogenic Ca-zeolites as soil modifiers along with gypsum, in making naturally degraded soils resilient and healthy.

Keywords $CaCO_3$ formation · Regressive pedogenesis · Remedial measures · SAT soils · Subsoil sodicity

6.1 Introduction

Soil is an integral part of land and land degradation is believed to be an anthropogenic induced process. The soil degradation is a big threat in maintaining sustainable livelihood security of farmers as it reduces the agricultural crop productivity (ICAR-NAAS 2010). Both land and soil degradation is induced by natural and human-induced processes, which are considered as chemical, physical or a biological phenomenon. Most published research on soil degradation in general

© Springer International Publishing AG 2017
D.K. Pal, *A Treatise of Indian and Tropical Soils*,
DOI 10.1007/978-3-319-49439-5_6

emphasizes the role of anthropogenic factors. One of the best examples is the development of saline and sodic soils in the north-west (NW) part of the IGP by introducing canal irrigation. The canal irrigation introduced during the end of the 19th century in the NW part of the IGP, is an important factor that affects the IGP soils. It was introduced to minimize the problem of aridity and to stabilize crop yields. Although it resulted in the expansion of the cultivable area in the IGP, the practice of irrigation during the dry climate without the provision of drainage led to soil salinization and alkalinisation within a few years, due to rise in the groundwater table containing high proportion of sodium relative to divalent cations and/or high residual alkalinity (Abrol 1982a). In addition, the application of groundwater with high sodicity for irrigation also increased sodic soils in the IGP (Abrol 1982b). Canal irrigation has caused similar salinity and sodicity problem also in shrink-swell and ferruginous soils of SAT environments. Even amidst such development of salinity and sodicity, recent research done in India during the last two decades has demonstrated the development of sodicity in soils in NW part of the IGP, shrink-swell soils of western and central India and ferruginous soils of southern India as the natural chemical soil degradation (Pal et al. 2016).

Physical degradation like water erosion is the most widespread form of soil degradation in India especially in the most affected RF soils of both HT and SAT climatic environments (ICAR-NAAS 2010). However, it has been demonstrated that the rate of top soil formation in Ultisols, Alfisols, Mollisols and Inceptisols of north eastern hills (NEH) (Bhattacharyya et al. 2007) and southern peninsular areas with clay enriched B-horizon is much higher than the soil loss by water erosion (Pal et al. 2014). Thus it is difficult to accept the higher rate of soil degradation in RF soils due to water erosion, and therefore demands a relook into the published literature.

In view of the background information, a proper identification of both natural and human-induced soil degradation in Indian soils is warranted. Such science backed exercise will help to identify the basic cause-effect relation of soil degradation in order to suggest the proper rehabilitation measures to make degraded soils resilient.

6.2   Natural Chemical Degradation of Soils in the Indian SAT, and Remedial Measures

Out of the 12 soil orders in Soil Taxonomy six diversified soil orders (Alfisols, Aridisols, Entisols, Inceptisols, Mollisols and Vertisols) do occur in the Indian semi-arid tropics (SAT) (Pal et al. 2000). In US Soil Taxonomy the SAT environments are identified at the suborder level within the ustic moisture regime. Although the ustic soil moister regime suggests the prevalence of dryness during a few months of the year, enough soil moisture is actually available for potentially growing crops in the rainy season (Soil Survey Staff 1975). The specific definition of ustic moisture regime is based on the mean annual soil temperature, and the duration of the period in which the control section of the profile remains moist or dry (Soil Survey Staff 1975). This definition means that the ustic moisture regime occurs in the tropical regions, with a monsoon climate that has at least one rainy season lasting three or more months in a year (Soil Survey Staff 1975).

Rain-fed agriculture with low productivity still prevails in the majority of the SAT soils. Although an increase in food production due to the implementation of improved soil, water and nutrient management practices is realized, many parts of the world continue to face food insecurity. Majority of the world's population ($\sim 60\%$) is facing food insecurity resides in South Asia and sub-Saharan Africa. Most of these areas are rain-fed and there are several challenges in improving the livelihoods of the rural poor. Ironically the rain-fed areas are mostly inhabited by poor rural communities. Moreover the rain-fed agriculture in the SAT area is fragile due to climatic variability in terms of spatial and temporal variation of rainfall. Besides, the rainfall is a short duration with high intensity phenomenon, which causes severe soil erosion, leading to physical, chemical and biological degradation of soils (Pal et al. 2000, 2012a, 2014; Srivastava et al. 2015).

The menace of degradation in SAT areas has made all scientists, agriculturists, environmentalists and policy makers anxious whether the soil resource base will be enough to feed the expected 8.2 billion world population by 2030 (www.unpopulation.org). In reality, soils are dynamic and are capable to supply nutrients, buffer acid-base reactions, absorb and degrade pathogens, detoxify and attenuate xenobiotic and inorganic compounds. Moreover, soils also possess the capacity of self-restoration through the process of soil formation. But soil formation is a slow process, and a substantial amount of soil can form only over a long geologic time. The misuse of soils and associated extreme climatic conditions can damage such self-regulating capacity and give way to regressive pedogenesis (Pal et al. 2013). This unfavourable pedogenesis might lead to the soil to regress from higher to lower usefulness and or drastically diminished productivity. Such an unfavourable endowment of soils is termed soil degradation (Lal et al. 1989).

6.3 Definition, Processes and Factors of Soil Degradation

Lal et al. (1989) defined soil degradation as diminution of soil quality (and thereby its current and potential productivity) and or a reduction in its ability to be a multi-purpose resource due to both natural and human-induced causes. The authors identified various processes that lead to soil degradation and they are accelerated erosion, increasing wetness and poor drainage, laterization, salinization, nutrient imbalance, decline in soil organic matter, and reduction in activity and species activity of soil fauna and flora. Processes of soil degradation are identified as chemical, physical and biological degradation of soils. The interactions among these factors affect the capacity of a soil for self-regulation and productivity (Lal et al. 1989).

Factors of soil degradation are both natural and human-induced in nature. These factors enhance degradation, leading to changes in properties of soils and the attributes for their life support (Lal et al. 1989). Some selected pedogenic processes such as laterization, hard setting, fragipan formation and clay-pan formation are hitherto considered as natural soil degradation processes as they lead to less desirable physical and chemical conditions of soils (Lal et al. 1989; Hall et al. 1982). But the majority of the information on soil degradation at national (Sehgal and Abrol 1992, 1994), regional (FAO 1994) or international level (Oldeman 1988; UNEP 1992) has focused only on degradation due to anthropogenic activities. Amidst such generalized statement a few recent reports on major soil types (Indo-Gangetic Plains or IGP, red ferruginous and deep black soils) confirm the development of sodicity and accumulation of relatively higher amounts of exchangeable Mg (EMP) than that of exchangeable Ca (ECP) in soils are also a natural process of soil degradation in semi-arid tropical (SAT) climatic conditions (Balpande et al. 1996; Pal et al. 2000, 2001, 2003a, b, 2006, 2012a, 2016; Vaidya and Pal 2002; Chandran et al. 2013).

6.4   Natural Chemical Soil Degradation: Neotectonic-Climate Linked and Mineral Induced

During the Quaternary especially in the last post-glacial period, the soils at many sites worldwide witnessed climatic fluctuations in response to the global climatic event, and the climatic changes were of frequent nature (Ritter 1996). Tectonic slopes and or faults determine the courses of large rivers (Singh et al. 2006) and play a significant role in the evolution of geomorphology and soils (Srivastava et al. 2015). The formation of the Western Ghats due to crustal movements also caused the change in climate from humid to semi-arid (Brunner 1970). FAO's (1994) endeavor to record land degradation in south Asia, the potential effects of global climatic change to cause soil degradation were not considered. But it was envisaged that if adverse changes occur in some areas, then these processes will certainly constitute a most serious form of human-induced degradation of natural resources. During the late Holocene period climate change from the humid to semi-arid did occur in major parts of the Indian subcontinent (Pal et al. 2009a, b, 2012a, 2014; Srivastava et al. 2015). Therefore, it is expected that the current aridic environment prevailing in many parts of the world including India might impair the physical and chemical properties of soils that may lead to reduced productivity of soils. Apart from the adverse effect of arid and semi-arid climates on soils' properties, the impairment of soil physical properties in presence of magnesium rich palygorskite minerals has also been observed (Pal et al. 2012a; Pal 2013). Thus a new research initiative is warranted to identify the changes in soil properties in the SAT due to climate change and Mg-minerals. Research database thereof can help in expanding our basic knowledge in pedology and provide opportunity to develop relevant database (Pal et al. 2009a). Such a database could be of immense value while adapting sustainable soil management and long-range resource management strategies for many developing nations in the arid and semi-arid regions of the world, especially in the Indian subcontinent, where arid and semi-arid environments cover more than 50% of the total geographical area (Pal et al. 2000). Considerable account on soil degradation due to anthropogenic activities (Sehgal and Abrol 1992, 1994; FAO 1994; Oldeman 1988; UNEP 1992) is available, which however fails to relate to an important issue of the natural chemical soil degradation that is causing reduction in soil productivity. Presently a precise account of factors of natural chemical degradation in major soil types of India is available, which forms a robust research database (Pal et al. 2012b, 2014; Srivastava et al. 2015; Bhattacharyya et al. 2004) to expand the current knowledge on natural soil degradation and to protect the livelihood of humankind.

6.5   Chemical Soil Degradation: Regressive Pedogenic Processes in the SAT

The SAT soils are calcareous, in general and the soils with mean annual rainfall (MAR) <800 mm are also sodic either in the subsoil or throughout the depth of soil profile (Pal et al. 2000, 2006). But it is to be noted that all calcareous soils are not sodic, but all sodic soils are calcareous (Pal et al. 2000). Calcareousness of soils is due to the presence of both pedogenic and non-pedogenic $CaCO_3$, but the pedogenic formation of $CaCO_3$ is not a favourable chemical reaction for soil health because this creates unfavourable physical conditions, caused by concomitant development of exchangeable sodium percent (ESP) (Pal et al. 2000; Balpande et al. 1996; Bhattacharyya et al. 2004). The presence of pedogenic $CaCO_3$ (PC) that is distinguished from the pedorelict $CaCO_3$ (NPC) by the soil thin section studies (Pal et al. 2000), is very common in major soil types of India (alluvial soils of the Indo-Gangetic plains, ferruginous soils and shrink-swell soils) (Fig. 6.1a, b; please refer to Fig. 3.6b). Water loss through evapo-transpiration is considered the primary mechanism in the precipitation of PC in the SAT environments, while temperature controls the water flow in the soil (Pal et al. 2000). Despite having low MAR (<500 mm) the development of sodicity is not common in the desert soil profiles (within ∼ 100 cm soil depth) due to their sandy textural class, ensuring better leaching of bicarbonates; and thus PC is generally observed at greater depth (Pal et al. 2000). However, in the loamy and clayey textured soils, the leaching of bicarbonates is slow and thus both PC and sodicity develop in upper horizons (Pal et al. 2000). Such pedogenetic processes are well exhibited in the SAT ferruginous soils (Alfisols) of southern India. These Alfisols developed in humid tropical climate of pre-Pliocene geological period, have restricted leaching because of ∼ 30%

(a)   (b)

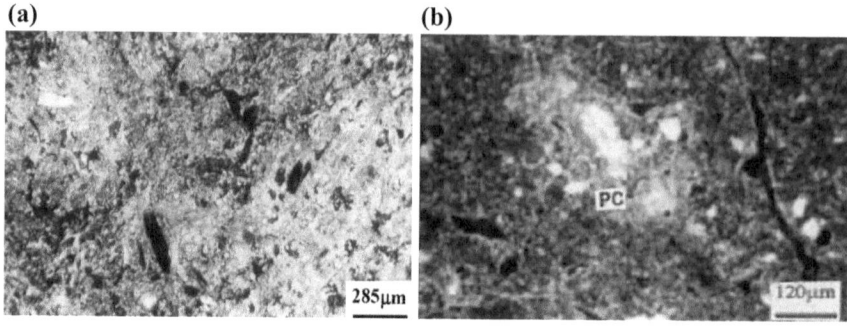

Fig. 6.1 Representative micromorphological features of pedogenic CaCO$_3$ (PC) in cross-polarized light **a** soils of the IGP (Natrustalfs), **b** Vertisols of central India (Adapted from Pal et al. 2009a)

clay, which are dominated by 2:1 expanding clay minerals. In these clayey and smectitic Alfisols, the formation of PC is observed due to the impact of the present day semi-arid climate, making these soils calcareous, unlike the ferruginous soils of humid tropical soils. The PCs in such soils with restricted leaching, are mainly concentrated as lubinites (please refer to Fig. 3.6b) that are formed only when the soil solution is supersaturated with CaCO$_3$ in the semi-arid environments (Wright 1988). This particular pedogenetic way of PC formation suggests that in addition to the climatic aridity, the texture has an important role in the accumulation of carbonates in soils (Wieder and Yaalon 1974).

The formation of PC in SAT soils as soil inorganic carbon (SIC) hitherto is considered an undesirable natural endowment because it impairs the soil productivity (Pal et al. 2000; Srivastava et al. 2002) and enhances the pH and also the relative abundance of Na$^+$ ions on soil exchange sites and in the solution. The Na$^+$ ions in turn cause dispersion of the fine clay particles. The dispersed fine clays translocate in major soil types of India (Pal et al. 2012a, 2014; Srivastava et al. 2015) as the formation of PC creates a Na$^+$—enriched chemical environment conducive for the deflocculation of clay particles and their subsequent movement downward. The formation of PC and the clay illuviation are thus two concurrent and contemporary pedogenetic events, resulting in increase in relative proportion of sodium, causing increased sodium adsorption ratio (SAR) and ESP and pH values with depth (please refer to Fig. 2.3; Fig. 6.2a, b). These pedogenetic processes represent a pedogenic threshold during the dry climates of the Holocene (Pal et al. 2012b, 2014, 2016; Srivastava et al. 2015), and clearly suggest that the formation of PC is a basic natural chemical degradation process (Pal et al. 2000, 2016), induced by tectonics-climate linked events (Pal et al. 2003b, 2009a, b), which exhibits the regressive pedogenesis (Pal et al. 2013, 2016); and it also immobilizes soil carbon (C) in unavailable form.

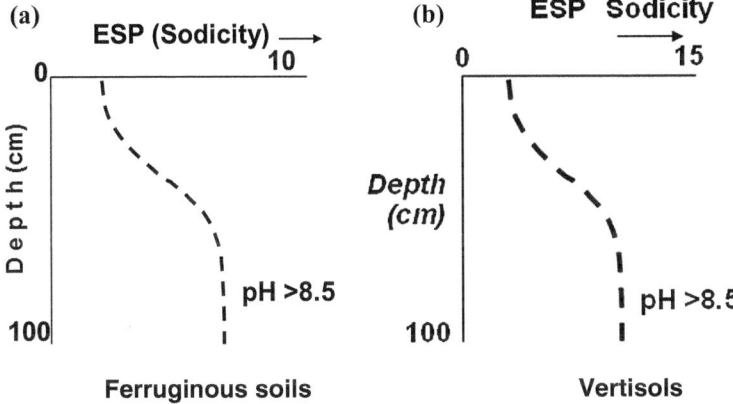

Fig. 6.2 Illuviation of Na-clay triggered by formation of PC, causing higher ESP and pH in the subsoils (Adapted from Pal et al. 2009a)

6.6   Clay Mineral (Palygorskite) Induced Natural Chemical Soil Degradation

Although the United States Salinity Laboratory (Richards 1954) grouped Ca^{2+} and Mg^{2+} ions together since both these ions improve the soil structure, Vertisols of dry climates of peninsular India have poor drainage due to clay dispersion caused by exchangeable magnesium (Vaidya and Pal 2002). Therefore, the saturation of Vertisols not only with Na^+ ions but also with Mg^{2+} ions blocks small pores in the soil (Pal et al. 2006). In other words, Mg^{2+} ions are less efficient than Ca^{2+} ions at flocculating soil colloids. The deflocculation of soil colloids is further enhanced even by low ESP (>5, <15), which reduces the saturated hydraulic conductivity (shC) to <5 mm h^{-1}, causing a >50% reduction in cotton yield (Pal et al. 2012a, b). In presence or absence of soil modifiers, dispersion of clay colloids as deflocculated colloids and impairment of the shC of Vertisols is generally an effect of ESP or exchangeable magnesium percent (EMP). In contrast, the shC of zeolitic Vertisols of the Marathwada region of Maharashtra is reduced to < 10 mm h^{-1}, although the soils are non-sodic (Typic Haplusterts, Zade 2007), neutral to mildly alkaline pH, with ESP <5, and EMP increasing with depth. Interestingly, in some pedons, the EMP is greater than ECP at depths below 50 cm. These soils contain palygorskite mainly in the silt and clay fractions (Zade 2007). Palygorskite minerals are present in Typic Haplusterts and also in Sodic Haplusterts/Sodic Calciusterts in association with Typic Haplusterts in India (Zade 2007) and elsewhere (Heidari et al. 2008). Palygorskite is the most magnesium-rich of the common clay minerals (Singer 2002). Neaman et al. (1999) examined the influence of clay mineralogy on disaggregation in some palygorskite-, smectite-, and kaolinite-containing soils with ESP <5 of the Jordan and Betshe'an valley in Israel. This mineral is the most disaggregated of the clay minerals, and its fibre does not remain associated with or within aggregates in soils and suspensions even when the soils are saturated with Ca^{2+} ions. Thus the deflocculated clay size palygorskite particles move downward in the profile preferentially over smectite and eventually clog the soil pores (Neaman and Singer 2004). Therefore, Vertisols with palygorskite content with high EMP lead to the dispersion of the clay colloids that form a 3D mesh in the soil matrix. Clay dispersion induced by this mineral, ultimately causes drainage problems when such soils are irrigated, presenting a predicament for crop production. In view of their poor drainage conditions and loss of productivity, non-sodic Vertisols (Typic Haplusterts) with palygorskite minerals should be better considered as naturally-degraded soils. It is a unique situation but poorly drained soils caused by palygorskite also occur in other parts of the world, and therefore a new initiative to classify this group of soils is warranted (Pal et al. 2012a, 2016; Pal 2013).

6.7   Micro-topography: A Unique Factor of Natural Soil Degradation in SAT

The SAT soils that are calcareous have sodicity in the subsoil or throughout the soil profile. But the degree of sodicity varies in the soil scape. In the IGP and area representing ferruginous soils highly sodic soils are formed on the micro-low (ML), whereas non-sodic/less sodic soils are developed on the micro-high (MH) positions (Fig. 6.3a, b) (Pal et al. 2003b; Chandran et al. 2013). In contrast, sodic Vertisols (Sodic Haplusterts) are developed in MH and non-sodic/less sodic Vertisols (Typic Haplusterts) on ML positions (please refer to Fig. 2.5) (Vaidya and Pal 2002). This follows from this that the soils in a micro-topographical sequence have contrasting chemical characteristics. The soils at MH positions are less alkaline in the IGP and highly alkaline in SAT Vertisols areas, respectively. Pal et al. (2003b) explained the formation of sodic soils in the IGP and observed that the micro-lows are repeatedly flooded with surface water during brief high-intensity showers, and so the soils are subject to cycles of wetting and drying. This provides a steady supply of alkalis by hydrolysis of feldspars, leading to precipitation of calcium carbonate at high pH, and due to the illuviation of Na-clay sodicity develops in the subsoils. This initially impairs the hydraulic properties of the subsoils and eventually leads to the development of relatively high ESP in the subsoils. It follows then that the SAT climate and topography interact to facilitate greater penetration of bicarbonate-rich water in ML than in MH positions. The sHC of the soils at ML position is almost nil in the subsurface layers due to higher amount of clay smectite and ESP. The SAT climate of the area induced precipitation of carbonates which in turn has increased Na^+ ion in the exchange complex (Pal et al. 2000). This chemical reaction is common in IGP, ferruginous and black soil regions of India wherein SAT climate induces the precipitation of calcium carbonates with a concomitant development of subsoil sodicity (Pal et al. 2000, 2003b, 2016; Balpande et al. 1996; Vaidya and Pal 2002; Chandran et al. 2013).

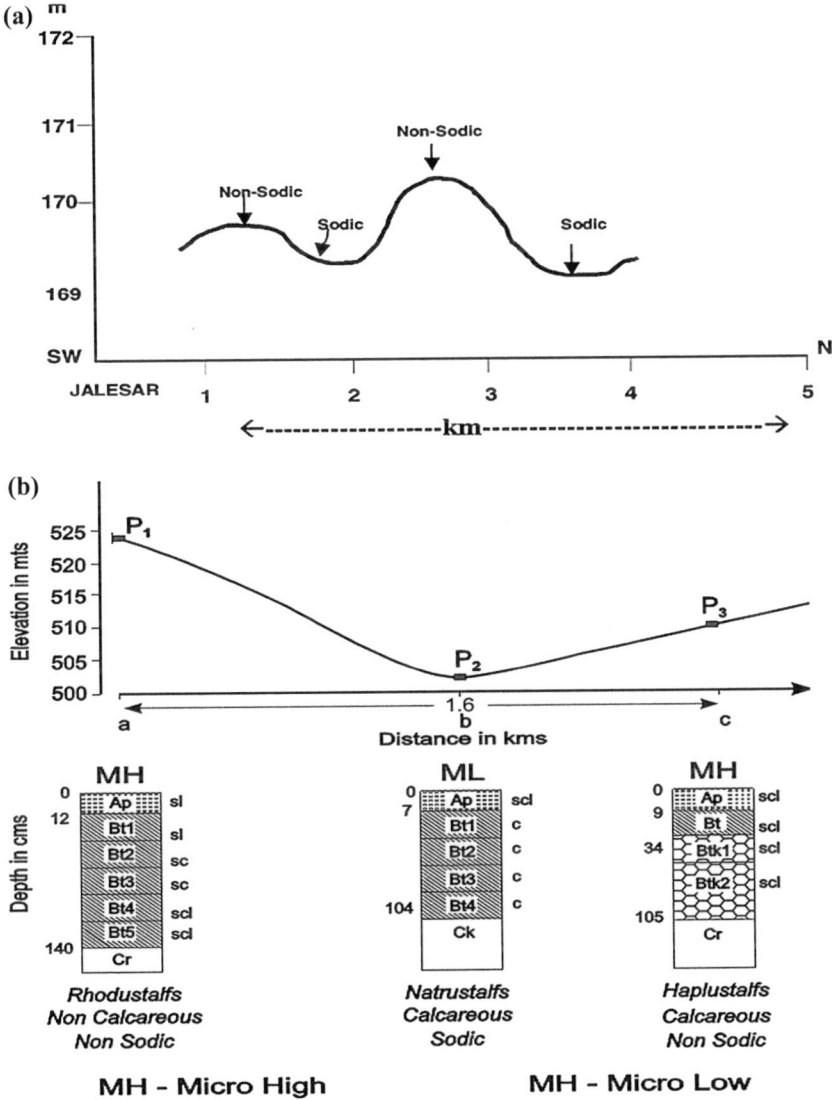

Fig. 6.3 a NE–SW profile in Ganga–Yamuna interfluve of the IGP showing non-sodic soils on MH and sodic soils on ML sites (Adapted from Pal et al. 2009a). **b** Schematic diagramme of the landscape representing the unique role of micro topography in the formation of non-sodic and sodic RF soils at MH and ML positions (Adapted from Chandran et al. 2013)

The formation of $CaCO_3$ enhances relatively more amount of Na^+ ions, which facilitates the illuviation of clay particles. Thus these two pedogenetic processes can be considered as the pedogenic processes occurring simultaneously as contemporary events in the drier climates in ML position. Similar micro-topographical situations in the formation of sodic and non-sodic soils on ML and MH positions respectively are observed in soils of the north-western parts of the IGP (Pal et al. 2003b) and also in ferruginous soils of southern India (Chandran et al. 2013). In contrast to this, in swell-shrink soils of central India a reverse situation was observed; sodic soils occur in MH and non-sodic soils as ML position. It is an example of unique situation in pedological parlance. It is observed that relatively higher amounts of PC and subsoil sodicity in MH Vertisols than in ML soils, suggesting that the formation of MH sodic soils is due to relatively more aridity on MH than the ML positions (Vaidya and Pal 2002). Thus the development of $CaCO_3$ and sodicity in the soils of ML and MH positions may be widespread in similar SAT areas of India. The rate of formation of PC for bench mark soils in Indian IGP, black (shrink-swell) soils and red ferruginous soils of SAT is estimated to be 129, 37.5 and 30 kg $CaCO_3$ ha^{-1} $year^{-1}$ respectively (Pal et al. 2000). At present the rate of formation of carbonates in ferruginous and black SAT soils is not alarming. But due care is needed while irrigating these soils especially those with clayey texture. In view of a high rate of $CaCO_3$ formation in the drier part (SAT) of the IGP soils, immediate remedial measures are required to make them resilient.

6.8   Indices of Soil Degradation

Research results indicate that the impairment of soil physical properties due to high ESP/EMP, and the presence of palygorskite mineral, as judged by reduced sHC, explains the cause-effect relation for the development of natural soil chemical degradation in SAT. In reality, such an explanation for the soil degradation remains elusive until it is related to crop performance and yield. Also, the critical limits of these indices have not been established. Sodicity tolerance ratings of crops in loamy-textured soils of the IGP indicated that a 50% reduction in relative yield of rice and wheat was observed when soil ESP was above 50 and ~ 40, respectively (Abrol and Fireman 1977). In shrink–swell soils (Vertisols), an optimum yield of cotton will be possible when soils are non-sodic (ESP < 5) and have sHC > 20 mm h^{-1}. About 50% reduction in yield occurs when soils are sodic (ESP > 5) showing low sHC (<10 mm h^{-1}). However, the Ca-rich zeolitic black soils (Sodic Haplusterts) of Rajasthan and Gujarat support rainfed crops fairly well (Pal et al. 2006, 2009c) due to favourable soil drainage (sHC > 10 mm h^{-1}). Therefore, fixing a lower limit of sodicity (Pal et al. 2006) at ESP > 40 for soils of the IGP (Abrol and Fireman 1977), at ESP >5 but <15 for Indian Vertisols (Balpande et al. 1996; Kadu et al. 2003), at ESP 6 for Australian soils or at ESP >15 for all soil types (Soil Survey Staff 1999) does not seem relevant for Vertisols (Pal et al. 2006). In view of the impairment of the hydraulic properties of soils in SAT a value of sHC < 10 mm h^{-1} (as weighted mean in 0–100 cm depth of soil) was advocated to define sodic soils instead of using an ESP or SAR (Pal et al. 2006). Hence, the deciding feature of soil classification must be the native vegetation because it indicates the nature of the land much more explicitly and authoritatively than any other arbitrary definition or nomenclature (Pal et al. 2000).

6.9 Rehabilitation of Degraded Soils

Research results obtained on the loss of Ca^{2+} ions from the soil system due to the formation of PC and concomitant development of sodicity indicate that SAT soils are prone to chemical degradation that creates strikingly different soil properties as compared to a normal soil. Therefore maintenance of required balance between exchangeable and water soluble Ca^{2+} ions in soil systems is of fundamental importance and thus poses a challenge for land resource managers to make these degraded soils resilient.

6.9.1 Difficulties in Identification and Reclamation of Sodic Soils

Sodic soils in the north-western (NW) parts of the IGP show salt-efflorescence in the surface as an evidence of soil sodicity, and this soil characteristic is mappable using remote sensing. Such maps for the shrink-swell soils (Vertisols and inter-grades) are not available due to general lack of salt-efflorescence in the soil surface. Another difficulty remains in the identification of sodicity in shrink-swell soils as they do not qualify as salt-affected soils as per the United States Salinity Laboratory criteria even when these soils have poor hydraulic properties of (<10 mm h^{-1}) at an ESP > 5 but < 15. The Central Soil Salinity Research Institute (ICAR), Karnal developed reclamation technology, which advises the use of mined gypsum, followed by paddy as the first crop (Abrol and Fireman 1977). The success of this technology however depends upon on the proper identification of nature of sodicity in all major soil types and also on the quantum of subsidy received by land holders (offered by the different Govt. and non-Govt. organization) of such problem soils for the procurement of the raw gypsum.

6.9.2 An Alternative Management to Reclaim Sodic Soils

Non-zeolitic and non- gypsiferrous Vertisols (Sodic Haplusterts) in SAT show poor shC (<10 mm h^{-1}) even at ESP $\geq 5 < 15$, and have poor crop productivity, and are impoverished in organic carbon (OC) but are rich in $CaCO_3$. In the long-term experiment such soils show enough resilience under improved management (IM) system of the ICRISAT (International Crops Research Institute for Semi-Arid Tropics) (Wani et al. 2003) that does not include any amendment like gypsum and FYM. Soil productivity is enhanced and the average grain yield of the IM system over thirty years was five times more than that in the traditional management (TM) system. Adaptation of the IM system improved physical, chemical and biological properties of soils to the extent that a poorly drained black soil (Sodic

Haplusterts) can now qualify for well-drained soil (Typic Haplusterts). A continuous release of higher amount of Ca^{2+} ions during the dissolution of $CaCO_3$ (8.4 mg/100 g soil/year in 1 m deep profile) under the IM system, compared to slower rate of formation of $CaCO_3$ (0.10 mg/100 g soil/year in 1 m profile), provide enough soluble Ca^{2+} ions to replace unfavourable Na^+ ions on the soil exchange sites. Higher exchangeable Ca/Mg ratio in soils under IM system improved the shC for better storage and release of soil water during dry spell between rains. Adequate supply of soil water helped in better crop productivity and also in higher OC sequestration. Such remarkable improvement in Vertisols' sustainability suggests that the IM system can mitigate the adverse effect of climate change (Pal et al. 2012b, c). The IM management protocol though very slow as compared to gypsum-aided one, is however a cost-effective and farmer–friendly. This technology is recommendable for a large scale impact on agricultural productivity (Wani et al. 2007). The above research results open up an interesting area of soil research that is to realize the benefit of the presence of $CaCO_3$ as a hidden treasure during the reclamation of sodic soils even with the addition of gypsum as practised in NW part of the IGP (Pal et al. 2009a, 2016). This has been detailed through C transfer mode (Bhattacharyya et al. 2004) which was reported to work better in the drier part of the IGP (Pal et al. 2009a).

6.9.3 Linking Calcium Carbonates to Resilience of the IGP Sodic Soils

Sodic soils (Natrustalfs) after following the reclamation protocol, show improvements in their morphological, physical and chemical properties so much that these soils can now reclassified as well-drained and OC-rich normal Alfisols (Haplustalfs). At present gypsum as amendment is added in relatively less amount than estimated by 'Gypsum Requirement' of highly sodic soils with ESP ~ 90–100 (Natrustalfs). But even with the low amount of added gypsum, sodic soils are reclaimed to show their resilience. Thus the success of this reclamation protocol cannot be fully credited to gypsum added because it does not enrich soil solution by the required amount of Ca^{2+} ions to replace Na ions on the soil exchange sites. Therefore, the Ca requirement to replace all exchangeable Na^+ ions is aided by Ca^{2+} ions available through the rapid dissolution of the native pedogenic $CaCO_3$ (PC) during the growing of the rice crop under submerged conditions. The rate of dissolution of PC during 30 months' cultural practice with gypsum in Natrustalfs is estimated to be 254 mg/100 g soil in the top 1 m of the profile. This indicates a much higher rate of dissolution (~ 100 mg/100 g soil/year) (Pal et al. 2009a) than its rate of formation (0.86 mg/100 g soil/year) in the top 1 m soil depth (Pal et al. 2000). It is worth mentioning here that the current theories favouring blue-green algae as a biological amendment to bring about sodic soil reclamation are untenable because of its inability to mobilize Ca from native $CaCO_3$ and thus are not comparable with an effective chemical amendment such as gypsum (Rao and Burns 1991).

6.9.4 Calcium Carbonates as Soil Modifier in Ensuring Soil Sustainability of SAT Soils

After becoming non-sodic in nature through cultural practices, both IGP soils (Haplustalfs) and Vertisols (Haplusterts) still remained with substantial stock of $CaCO_3$ (SIC) in the first 1.5 m depth, which has potential to improve the drainage, establishment of vegetation, and also sequestration of OC in soils (Pal et al. 2012c). The continuance of agronomic practices of the NARS (National Agricultural Research Systems) and ICRISAT can provide the most important Ca^{2+} ions both in solution and exchange sites of soil. These resilient soils still contain nearly 2–7% $CaCO_3$. In view of the rate of its dissolution (~ 100 mg/100 g soil/year for IGP soils, and 21 mg/100 g soil/year for Vertisols, Pal et al. 2012a, c), it is envisaged that under improved management in the SAT environment, total dissolution of $CaCO_3$ would take a time of couple of centuries. Such chemical environment enriched with Ca^{2+} ions would not allow both Haplustalfs and Haplusterts to transform to any other soil order so long $CaCO_3$ would continue to act as a soil modifier (Pal et al. 2012a). Positive role of $CaCO_3$ in both reclamation and sequestration of OC in SAT soils may benefit maintaining the soil health of the farmlands if additional financial support through national and international initiatives including the incentives or transferable C credits under CDM is made available (Pal et al. 2015, 2016) to stake holders of sodic soils.

6.10 Nature and Extent of Degradation in RF Soils: A Critique

Water erosion is the most widespread form of soil degradation in the Indian sub-continent. It is estimated that the erosion affects ~ 73.3 m ha area spreading in all agro-climatic zones. The extent and severity of soil degradation are related to the intensity of the rainfall, slope of the land, and the types of soil and land use. Soils under HT climate of NEH and southern peninsular areas are most affected by water erosion, and the highest area under this category of degradation is in Nagaland (87%), followed by Meghalaya (78%), Arunachal Pradesh (73%), Assam (66%), Manipur (53%), Tripura (38%), Sikkim (37%) and Kerala (15%) (ICAR-NAAS 2010).

It is worth mentioning here that the above estimates were based on an assumption that soil erosion <10 t ha^{-1} year^{-1} (using the empirical Universal Soil Loss Equation, USLE to estimate spatial variations of soil loss factors like R, K, L, S, C and P factors) generally does not significantly affect soil productivity; and the soils with loss <10 t ha^{-1} year^{-1} were not considered degraded (ICAR-NAAS 2010). Such assumption on soil loss appears to be far from real situation in soils under HT climate, because of the dominance of Ultisols, Alfisols, Mollisols, and Inceptisols with clay enriched B horizons in NEH areas, Kerala, Goa, Maharashtra,

Madhya Pradesh, Karnataka and Andaman and Nicobar Islands (please refer to Table 3.1). In the event of such amount of soil loss persistence of these soil orders with clay enriched B horizons would not have been a reality. Moreover, formation and persistence of such soil orders are possible when landscape attains the stability.

The major pedogenetic processes that are operative to give rise to these soil orders, are evident through the addition of C by litter falls and its accumulation as soil organic matter under adequate vegetation and climate, translocation of clay particles (to form clay enriched B horizons) and transformation of 1.4 minerals to kaolin (detailed in Chap. 3). The rate of soil formation varies from <0.25 mm year^{-1} in dry and cold environments to >1.5 mm year^{-1} in humid and warm environments (Kassam et al. 1992).

If the rate of soil formation is taken as 2.0 mm year^{-1} for soils of HT climate, the amount of top soil formation would be around 29 t ha^{-1} year^{-1} (as shown for soils of Tripura in NEH, Bhattacharyya et al. 2007). This gain in soil clearly suggests that the rate of soil loss by water erosion from Ultisols, Alfisols, Mollisols and Inceptisols is minimal (Bhattacharyya et al. 2007). The pedogenetic processes in soils of HT climate ensures the positive balance of soil formation, and thus mature soils (like Ultisols, Alfisols, Mollisols and clay enriched Inceptisols) on a stable landscape under HT climate is a reality. In contrast, in RF soils (mainly Entisols) on higher slopes (ridges, scarps and terraces) under low vegetation with only shrubs and bushes, soil development is greatly hampered by the severe soil loss due to water erosion. Soil loss is also evident in other soils that have less vegetative cover. Besides proper mechanical conservation measures, such soils areas may be suitable for forestry, horticultural and plantation crops to build resilience in them (Bhattacharyya et al. 1998, 2007).

As per report of ICAR-NAAS (2010), the RF soils of Indian states under SAT environments also suffer soil loss due to erosion, and the loss is maximum in Karnataka (49%) followed by Andhra Pradesh (40%) and Tamil Nadu (20%), considering soil loss >10 t ha^{-1} year^{-1} as the threshold for soil degradation. It is to be noted that the RF soils of SAT dominantly belong to Alfisols, and the other soil orders are Inceptisols, Entisols and Mollisols (please refer to Table 3.1). If the rate of soil formation in dry environments is considered at <0.25 mm year^{-1} (Kassam et al. 1992), SAT Alfisols would gain soil at least 3.67 t ha^{-1} year^{-1}. This value is close to soil loss as evident from the results of short-term hydrological studies on small agricultural watershed on Alfisols at the ICRISAT Center, Patancheru, India, which indicate an average soil loss from SAT Alfisols under traditional system is around 3.84 t ha^{-1} year^{-1} (Pathak et al. 1987). On the other hand, the results from long-term study reported an annual soil loss of 4.62 t ha^{-1} (Pathak et al. 2013). Soil loss from both the ICRISAT experiments is much less than the soil loss value assumed by ICAR-NAAS (2010).

The higher soil loss on the SAT Alfisols under traditional management was attributed to crusting, sealing and low structural stability (Pathak et al. 1987, 2013). But many SAT Alfisols have clay enriched B horizons with substantial amount of smectite clay mineral in the subsoils (please refer to Fig. 3.2b) (Pal et al. 1989; Chandran et al. 2009), which causes moderate shrink-swell properties (coefficient of linear extensibility, COLE >0.06, Pal et al. 2014). Additionally, these Alfisols have reduced saturated hydraulic conductivity in the subsoils (Pathak et al. 2013). Such physical and mineralogical characteristics in the subsoil restrict vertical movement of water in the soil profile, resulting in greater soil loss from the SAT Alfisols through overland lateral flow of water. Vertical movement of water will be further restricted in SAT Alfisols that have subsoil sodicity due to the formation of pedogenic $CaCO_3$ (Pal et al. 2013).

In order to reduce unwarranted soil loss and also to improve the sustainability of the SAT Alfisols, improved system of management developed by the ICRISAT (Pathak et al. 1987) may help. ICRISAT's improved system (improved water management, land development and soil conservation practices combined with appropriate cropping systems, Pathak et al. 1987) minimized soil loss to nearly 1 t ha^{-1} year^{-1} and simultaneously increased crop productivity compared to traditional system. These research results aptly suggest that a threshold value of soil loss by water erosion as a sign of degradation, must be based on experimental results rather than assuming an arbitrary value of >10 t ha^{-1} year^{-1} (ICAR-NAAS 2010).

Development of acidity in soils is indeed a sign of chemical degradation. Acid soils develop under HT climatic environment with a loss of soil fertility. While estimating the area degraded by acidity, soils with strong (pH < 4.5) and moderate acidity (pH 4.5–5.5) only were considered. With this assumption, about 6.98 m ha area is affected by soil acidity, which is about 9.4% of the total geographical area of the country (ICAR-NAAS 2010). Soils of HT climate in the states of Kerala, Goa, Karnataka, Tamil Nadu, Maharashtra and NEH areas are strong to moderately acidic Alfisols, Ultisols and Mollisols (please refer to Table 3.2); and their further weathering in HT climate would finally close at Ultisols with considerable amount of layer silicate minerals (Pal et al. 2012a, 2014). Such OC rich acid soils do respond to management interventions and support luxuriant forest vegetation, horticultural, cereal crops, tea, coffee and spices (Sehgal 1998). In view of such reality, it would not be prudent to class them as chemically degraded soils.

It is to be noted that at present the extent of degradation (soil loss and soil acidity) of RF soils (especially Ultisols, Alfisols, Mollisols and Inceptisols with clay enriched B horizons) is not at an alarming stage, which is in contrast to what so far has been projected by assuming arbitrary values for soil loss. But they require improved management (IM) practices developed by researchers to upgrade and maintain their nutrient status and make efficient use of soil water to sustain crop productivity at an enhanced level. The IM system package available includes improved seeds, NPK fertilizers, micronutrients, FYM and use of gypsum (for sodic soils), use of legumes in the cropping sequence (Wani et al. 2003), lime (Datta 2013), and improved water management, land development and the implementation of the soil conservation practices for acidic Alfisols, Ultisols and Inceptisols (Datta 2013) and SAT Alfisols (Pathak et al. 1987, 2009, 2013).

References

Abrol IP (1982a) Reclamation of waste lands and world food prospects. In: Whither soil research, panel discussion papers, 12th International Congress Soil Science, New Delhi, pp 317–337

Abrol IP (1982b) Reclamation and management of salt-affected soils. In: Review of soil research in India. Part 11, 12th International Congress Soil Science, New Delhi, pp 635–654

Abrol IP, Fireman M (1977) Alkali and Saline soils; identification and improvement for crop production. Bulletin No. 4. Central Soil Salinity Research Institute, Karnal, India

Balpande SS, Deshpande SB, Pal DK (1996) Factors and processes of soil degradation in Vertisols of the Purna valley, Maharashtra, India. Land Degrad Dev 7:313–324

Bhattacharyya T, Mukhopadhyay S, Baruah U, Chamuah GS (1998) Need for soil study to determine degradation and landscape stability. Curr Sci 74:42–47

Bhattacharyya T, Pal DK, Chandran P, Mandal C, Ray SK, Gupta RK, Gajbhiye KS (2004) Managing Soil carbon stocks in the Indo-Gangetic Plains, India, Rice-Wheat Consortium for the Indo-Gangetic Plains, New Delhi—110 012, India, p 44

Bhattacharyya T, Babu R, Sarkar D, Mandal C, Dhyani BL, Nagar AP (2007) Soil loss and crop productivity model in humid tropical India. Curr Sci 93:1397–1403

Brunner H (1970) Pleisitozane klimaschwankengen im Bereich den ostlichen Mysore-Plateaus (Sudindien). Geologie 19:72–82

Chandran P, Ray SK, Durge SL, Raja P, Nimkar AM, Bhattacharyya T, Pal DK (2009) Scope of horticultural land-use system in enhancing carbon sequestration in ferruginous soils of the semi-arid tropics. Curr Sci 97:1039–1046

Chandran P, Ray SK, Bhattacharyya T, Tiwari P, Sarkar D, Pal DK, Mandal C, Nimkar A, Maurya UK, Anantwar SG, Karthikeyan K, Dongare VT (2013) Calcareousness and subsoil sodicity in ferruginous Alfisols of southern India: an evidence of climate shift. Clay Res 32:114–126

Datta M (2013) Soils of north-eastern region and their management for rain-fed crops. In: Bhattacharyya T, Pal DK, Sarkar D, Wani SP (eds) Climate change and agriculture. Studium Press, New Delhi, pp 19–50

FAO (Food and Agriculture Organization of the United Nations) (1994) Land degradation in South Asia: its severity, causes and effects upon the people. World Soil Resources Reports 78, FAO, Rome, Italy

Hall GF, Daniels RB, Foss JE (1982) Rates of soil formation and renewal in the USA. In: Schmidt BL (ed) Determinants of soil loss tolerance. ASA Publication No. 45. American Society of Agronomy, Madison, Wisconsin, USA, pp 23–29

Heidari A, Mahmoodi Sh, Roozitalab MH, Mermut AR (2008) Diversity of clay minerals in the Vertisols of three different climatic regions in western Iran. J Agric Sci Technol 10:269–284

ICAR-NAAS (Indian Council of Agricultural Research- National Academy of Agricultural Sciences) (2010) Degraded and waste lands of India—status and spatial distribution. ICAR-NAAS. Published by the Indian Council of Agricultural Research, New Delhi, 56 pp

Kadu PR, Vaidya PH, Balpande SS, Satyavathi PLA, Pal DK (2003) Use of hydraulic conductivity to evaluate the suitability of Vertisols for deep-rooted crops in semi-arid parts of central India. Soil Use Manage 19:208–216

Kassam AH, van Velthuizen GW, Fischer GW, Shah MM (1992) Agro-ecological land resource assessment for agricultural development planning: a case study of Kenya resources data base and land productivity. Land and Water Development Division, Food and Agriculture Organisation of the United Nations and International Institute for Applied System Analysis, Rome

Lal R, Hall GF, Miller FP (1989) Soil degradation. I. Basic processes. Land Degrad Rehabil 1:51–69

Neaman A, Singer A (2004) The effects of palygorskite on chemical and physico–chemical properties of soils: a review. Geoderma 123:297–303

Neaman A, Singer A, Stahr K (1999) Clay mineralogy as affecting disaggregation in some palygorskite-containing soils of the Jordan and Bet-She'an Valleys. Aust J Soil Res 37:913–928

Oldeman LR (ed) (1988) Global assessment of soil degradation (GLASOD). Guidelines for general assessment of status of human-induced soil degradation. ISRIC, Wageningen, the Netherlands

Pal DK (2013) Soil modifiers: their advantages and challenges. Clay Res 32:91–101

Pal DK, Deshpande SB, Venugopal KR, Kalbande AR (1989) Formation of di and trioctahedral smectite as an evidence for paleoclimatic changes in southern and central Peninsular India. Geoderma 45:175–184

Pal DK, Dasog GS, Vadivelu S, Ahuja RL, Bhattacharyya T (2000) Secondary calcium carbonate in soils of arid and semi-arid regions of India. In: Lal R et al (eds) Global climate change and pedogenic carbonates. Lewis Publishers, FL, USA, pp 149–185

Pal DK, Balpande SS, Srivastava P (2001) Polygenetic vertisols of the Purna Valley of central India. Catena 43:231–249

Pal DK, Srivastava P, Bhattacharyya T (2003a) Clay illuviation in calcareous soils of the semi-arid part of the Indo-Gangetic Plains, India. Geoderma 115:177–192

Pal DK, Srivastava P, Durge SL, Bhattacharyya T (2003b) Role of micro topography in the formation of sodic soils in the semi-arid part of the Indo-Gangetic Plains, India. Catena 51:3–31

Pal DK, Bhattacharyya T, Ray SK, Chandran P, Srivastava P, Durge SL, Bhuse SR (2006) Significance of soil modifiers (Ca-zeolites and gypsum) in naturally degraded vertisols of the Peninsular India in redefining the sodic soils. Geoderma 136:210–228

Pal DK, Bhattacharyya T, Chandran P, Ray SK (2009a) Tectonics-climate linked natural soil degradation and its impact in rainfed agriculture: Indian experience. In: Wani SP et al (eds) Rainfed agriculture: unlocking the potential. CAB International Publishing, Oxfordshire, pp 54–72

Pal DK, Mandal DK, Bhattacharyya T, Mandal C, Sarkar D (2009b) Revisiting the agro-ecological zones for crop evaluation. Indian J Genet 69:315–318

Pal DK, Bhattacharyya T, Chandran P, Ray SK, Satyavathi PLA, Durge SL, Raja P, Maurya UK (2009c) Vertisols (cracking clay soils) in a climosequence of Peninsular India: evidence for holocene climate changes. Quatern Int 209:6–21

Pal DK, Wani SP, Sahrawat KL (2012a) Vertisols of tropical Indian environments: pedology and edaphology. Geoderma 189–190:28–49

Pal DK, Bhattacharyya T, Wani SP (2012b) Formation and management of cracking clay soils (vertisols) to enhance crop productivity: Indian experience. In: Lal R, Stewart BA (eds) World soil resources. Francis and Taylor, pp 317–343

Pal DK, Wani SP, Sahrawat KL (2012c) Role of calcium carbonate minerals in improving sustainability of degraded cracking clay soils (Sodic Haplusterts) by improved management: an appraisal of results from the semi-arid zones of India. Clay Res 31:94–108

Pal DK, Sarkar D, Bhattacharyya T, Datta SC, Chandran P, Ray SK (2013) Impact of climate change in soils of semi-arid tropics (SAT). In: Bhattacharyya et al (ed) Climate change and agriculture. Studium Press, New Delhi, pp 113–121

Pal DK, Wani SP, Sahrawat KL, Srivastava P (2014) Red ferruginous soils of tropical Indian environments: a review of the pedogenic processes and its implications for edaphology. Catena 121:260–278. doi:10.1016/j.catena2014.05.023

Pal DK, Wani SP, Sahrawat KL (2015) Carbon sequestration in Indian soils: present status and the potential. Proc Natl Acad Sci Biol Sci (NASB) India, 85:337–358. doi:10.1007/s40011-014-0351-6

Pal DK, Bhattacharyya T, Sahrawat KL, Wani SP (2016) Natural chemical degradation of soils in the Indian semi-arid tropics and remedial measures. Curr Sci 110:1675–1682

Pathak P, Singh S, Sudi R (1987) Soil and water management alternatives for increased productivity on SAT Alfisols. Soil conservation and productivity. In: Proceedings IV international conference on soil conservation, Maracay-Venezuela, 3–9 Nov 3–9 1985, pp 533–550

Pathak P, Sahrawat KL, Wani SP, Sachan RC, Sudi R (2009) Opportunities for water harvesting and supplemental irrigation for improving rainfed agriculture in semiarid areas. In: Wani SP,

Rockström J, Oweis T (eds) Rainfed agriculture: unlocking the potential. CABI International, Oxfordshire, pp 197–221

Pathak P, Sudi R, Wani SP, Sahrawat KL (2013) Hydrological behaviour of Alfisols and Vertisols in the semi-arid zone: implications for soil and water management. Agric Water Manage 118:12–21

Rao DLN, Burns RG (1991) The influence of blue-green algae on the biological amelioration of alkali soils. Biol Fertil Soils 11:306–312

Richards LA (ed) (1954) Diagnosis and improvement of saline and alkali soils. USDA Agriculture Handbook 60. USDA, Washington, DC, USA

Ritter DF (1996) Is quaternary geology ready for the future? Geomorphology 16:273–276

Sehgal JL (1998) Red and lateritic soils: an overview. In: Sehgal J, Blum WE, Gajbhiye KS (eds) Red and lateritic soils. Managing red and lateritic soils for sustainable agriculture 1. Oxford and IBH Publishing Co. Pvt. Ltd., New Delhi, pp 3–10

Sehgal JL, Abrol IP (1992) Land degradation status: India. Desertification Control Bull 21:24–31

Sehgal JL, Abrol IP (1994) Soil degradation in India: status and impact. Oxford & IBH Publishing Co. Pvt. Ltd, New Delhi

Singh S, Parkash B, Rao MS, Arora M, Bhosle B (2006) Geomorphology, pedology and sedimentology of the Deoha/Ganga-Ghaghara Interfluve, Upper Gangetic Plains (Himalayan Foreland Basin)—extensional tectonic implications. Catena 67:183–203

Singer A (2002) Palygorskite and sepiolite. In: Dixon JB, Schulze DG (eds) Soil mineralogy with environmental applications, SSSA Book Series, vol 7., Soil science society of AmericaMadison, WI, pp 555–583

Soil Survey Staff (1975) Soil taxonomy: a basic system of soil classification for making and interpreting soil surveys. United States Department of Agriculture Handbook No. 436. U.S. Government Printing Office, Washington, DC

Soil Survey Staff (1999) Soil taxonomy: a basic system of soil classification for making and interpreting soil surveys. United States Department of Agriculture, Natural Resource Conservation Service, Agriculture Handbook No. 436, U.S. Government Printing Office, Washington, DC

Srivastava P, Bhattacharyya T, Pal DK (2002) Significance of the formation of calcium carbonate minerals in the pedogenesis and management of cracking clay soils (Vertisols) of India. Clays Clay Miner 50:111–126

Srivastava P, Pal DK, Aruche KM, Wani SP, Sahrawat KL (2015) Soils of the Indo-Gangetic plains: a pedogenic response to landscape stability, climatic variability and anthropogenic activity during the holocene. Earth Sci Rev 140:54–71. doi:10.1016/j.earscirev.2014.10.010

UNEP (1992) World atlas of desertification. Edward Arnold, London

Vaidya PH, Pal DK (2002) Micro topography as a factor in the degradation of vertisols in central India. Land Degrad Dev 13:429–445

Wani SP, Pathak P, Jangawad LS, Eswaran H, Singh P (2003) Improved management of vertisols in the semi-arid tropics for increased productivity and soil carbon sequestration. Soil Use Manag 19:217–222

Wani SP, Sahrawat KL, Sreedevi TK, Bhattacharyya T, Srinivas Rao Ch (2007) Carbon sequestration in the semi-arid tropics for improving livelihoods. Int J Environ Stud 64:719–727

Wieder M, Yaalon DH (1974) Effect of matrix composition on carbonate nodule crystallization. Geoderma 43:95–121

Wright VP (1988) Pleokarsts and paleosols as indicators of paleoclimate and porosity evolution: a case study from the carboniferous of South Wales. In: James NP, Choquette PW (eds) Springer, New York, pp 329–341

Zade SP (2007) Pedogenic studies of some deep shrink-swell soils of Marathwada region of Maharashtra to develop a viable land use plan. Ph.D Thesis, Dr. P D K V, Akola, Maharashtra, India

Chapter 7
Clay and Other Minerals in Soils and Sediments as Evidence of Climate Change

Abstract Identification of paleoclimatic signatures in paleosols is a major challenge to soil scientists. It is realized that the clay minerals of the paleosols are potential promising materials for documenting and resolving a wide spectrum of different genetic environments and reactions. It is often difficult to determine which soil minerals are characteristic of different climatic zones as the environment itself changes over time with consequent further modification of mineral assemblage and this is particularly true for clay minerals. Clay minerals such as kaolinite often remain unaltered through subsequent changes in climate, and therefore, may preserve a paleoclimatic record. Other layered silicates at a less advanced stage of weathering may adjust to subsequent environmental changes and thus may lose their interpretative value for paleoclimatic signatures. However, Indian soil scientists, clay mineralogists and earth scientists indicate that minerals of intermediate weathering stage can act as potential indicators of paleoclimatic changes in parts of central India and Gangetic Plains. They have demonstrated how secondary minerals like di- and trioctahedral smectites (DSm and TSm), smectite-kaolinite interstratified minerals (Sm/K), hydroxy-interlayered smectite (HIS), hydroxy-interlayered vermiculite (HIV), pseudo-chlorite (PCh) of intermediate weathering stage, and $CaCO_3$ of pedogenic (PC) and non-pedogenic (NPC) origin can be regarded as potential indicators of paleoclimatic changes in major soil types of India and also in paleosols of the alluvial sediments of the Himalayan river systems and Cratonic source from Peninsular India.

Keywords Clay minerals · Soils · Sediments · Climate change · Polygenetic soils

7.1 Introduction

Perhaps majority of soils contain elements, which formed as a result of environmental conditions that have now altered. Paleosols have become standard tools in the study of Quaternary sequences and such studies are well documented. Paleosols are also abundant in the pre-Quaternary geological record and even Precambrian

© Springer International Publishing AG 2017
D.K. Pal, *A Treatise of Indian and Tropical Soils*,
DOI 10.1007/978-3-319-49439-5_7

types are now widely recognized. Paleopedological research unravels the signatures of climate change that generally remain stored in soils and sediments of the past (Pal et al. 2000a) and such soils are known as paleosols, formed on a landscape of the past (Valentine and Dalrymple 1976). The study of paleosols is still in infancy but studies on paleosols have caught the attention of the pedologists, sedimentologists and soil mineralogists in India and abroad (Singer 1980; Beckmann 1984; Jenkins 1985; Fenwick 1985; Wright 1986; Pal et al. 1989, 2001, 2009, 2012a; Srivastava et al. 1998, 2007, 2009, 2010, 2013).

Identification of paleoclimatic signatures in paleosols is a major challenge to soil scientists. Yaalon (1971) points to the fact that the products of the self-terminating, irreversible reactions such as calcareous or siliceous incrustations are among the most permanent and best indicators of paleo-environmental conditions. Whenever reliable paleoclimatic indicators, such as paleontological remains, pollen or isotope chemistry were absent or have failed, paleoclimatologists have turned to paleosols for clues as to the nature of climates of the past (Singer 1980). It is realized that the clay minerals of the paleosols are potential promising materials for documenting and resolving a wide spectrum of different genetic environments and reactions (Keller 1970). The use of clay minerals in paleosols and saprolites (weathering profiles) for the purpose of paleoclimatic interpretation is based on five assumptions (Singer 1980) and they are (i) clay minerals and climatic parameters (rainfall and temperature) are quantitatively related, (ii) clay minerals are stable and do not change as long as the climate remains stable and tectonic rejuvenation does not take place, (iii) clay mineral assemblages are uniform throughout the weathering profile, (iv) clay minerals are stable (post-burial stability) and (v) clay minerals have a uniform sensitivity towards environmental change. Singer (1980) further stated that since only soils are in direct contact with atmospheric agencies, the impact of climatic variables on soil clay formation is also most direct.

It is well known that equilibria are only apparent and in any case ephemeral in nature. Therefore, it is often difficult to determine which soil minerals are characteristic of different climatic zones. However, those clay minerals which occur most frequently can be considered to have climatic significance (Tardy et al. 1973). Environment itself changes over time with consequent further modification of mineral assemblage and this is particularly true for clay minerals (Jenkins 1985). Clay minerals such as kaolinite often remain unaltered through subsequent changes in climate, and therefore, may preserve a paleoclimatic record. Other layered silicates at a less advanced stage of weathering may adjust to subsequent environmental changes and thus may lose their interpretative value for paleoclimatic signatures (Singer 1980). However, recent research records of Indian researchers indicate that minerals of intermediate weathering stage can act as potential indicators of paleoclimatic changes in parts of central India and Gangetic Plains (Pal et al. 1989, 2009, 2012a; Srivastava et al. 1998, 2013). These authors have demonstrated how secondary minerals like di- and trioctahedral smectites (DSm and TSm), smectite-kaolinite interstratified minerals (Sm/K), hydroxy-interlayered smectite (HIS), hydroxy-interlayered vermiculite (HIV), pseudo-chlorite (PCh) of intermediate weathering stage, and $CaCO_3$ of pedogenic (PC) and non-pedogenic

(NPC) origin can be regarded as potential indicators of paleoclimatic changes in major soil types of India and also in paleosols of the alluvial sediments of the Himalayan river systems and Cratonic source from Peninsular India.

7.2 Di- and Tri-Octahedral Smectite as Evidence for Paleoclimatic Changes

Pal et al. (1989) reported the presence of fairly well crystallized dioctahedral smectites as the first weathering product of Peninsular Gneiss, which partly transformed to kaolin (interstratified hydroxy interlayered smectite and kaolinite, KI-HIS) in ferruginous soils (Alfisols) formed in a pre-Pliocene tropical humid climate. Careful examination confirms that such kaolin is not a discrete kaolinite as XRD diagrams of its Ca-saturated and glycolated sample indicates the broad base of 0.72 nm peak and tails towards the low angle. On heating the K-saturated sample at 550 °C, the 0.72 nm peak disappears, confirming the presence of kaolin and simultaneously reinforces the 1.0 nm region at much higher degree even in presence of 1.4 nm minerals. Thus it confirms the presence of KI-HIV/HIS (kaolinite interstratified with hydroxy-interlayered vermiculite, HIV or smectite, HIS) (please refer to Fig. 3.4b). When the termination of humid climate during the Plio-Pleistocene transition did occur, both these clay minerals were preserved to the present. Therefore, the ferruginous Alfisols overlying the saprolites dominated either by dioctahedral smectite or kaolin qualify to be relict paleosols (Pal et al. 1989; Chandran et al. 2000). During the Plio-Pleistocene transition period, such paleosols have been affected by the climatic change from humid to drier conditions. Effect of such climate shift is evidenced by the formation of trioctahedral smectite in the present dry climate from the sand and silt size biotite (please refer to Fig. 3.6a), which survived weathering during the earlier humid climate. This trioctahedral smectite is truly high charge smectite or low charge vermiculite that expands to 1.7 nm on glycolation of Ca-saturated sample but contracts readily to 1.0 nm on K-saturation and heating to 110 °C. The presence of both di- and trioctahedral smectite in relict paleosols is common as reported from ferruginous Alfisols of southern India (Pal et al. 1989; Chandran et al. 2000). This is a unique combination in soils, which helps in identifying the climate change in the geological past. The present day warm semi-aridic climatic conditions also favoured the formation of pedogenic calcium carbonate (PC) (please refer to Fig. 3.6b) by inducing the precipitation of $CaCO_3$ with a concomitant development of subsoil sodicity (Pal et al. 2000b, 2012a). These relict paleosols did preserve the dioctahedral smectite of earlier humid climate, and trioctahedral smectite and PC of the present semi-arid climate. Therefore, these relict paleosols qualify to be polygenetic soils with strong paleoclimatic potential (Pal et al. 1989).

7.3   Red and Black Soils in Semi-arid Climatic Environments

In semi-arid region of southern Peninsular India, the occurrence of spatially asso-ciated red ferruginous (Alfisols) and black (Vertisols) soils on gneiss under similar topographical conditions is very common (Pal and Deshpande 1987). Ferruginous soil clays consist mainly of kaolin and smectite whereas the low charge dioctahe-dral smectite is dominant in black soil clays. The inverse relation between kaolin and smectite with pedon depth of ferruginous soil clays (please refer to Fig. 3.2b) (Pal et al. 1989) indicated the transformation of smectite to kaolin but such transformation is not feasible in slightly acid to moderately alkaline soil reaction, which is a characteristic of the prevailing semi-arid climate. Moreover, the arid climate cannot yield the huge amount of smectite required for the formation of black soils (Vertisols) (Bhattacharyya et al. 1993). Earlier studies in southern Peninsular India (Murali et al. 1978; Rengasamy et al. 1978) suggested that kaolinite was formed in an earlier geological period with more rainfall and great fluctuations in temperature, as evidenced by the presence of granitic tors all around such area (Pal and Deshpande 1987). Therefore, the smectite of Vertisols formed in the earlier humid climate, which was detached from the weathering gneissic rock and transported downstream and deposited in low-lying areas following the land-scape reduction process (Bhattacharyya et al. 1993). Due to this process the typical Vertisols were developed in the micro depressions (please refer to Fig. 2.4). After the peneplanation, red ferruginous soils on stable surface continued to weather to form kaolin mineral as the stability of the smectite was ephemeral in tropical humid climate (Bhattacharyya et al. 1993). Termination of the humid climate did occur during the Plio-Pleistocene transition, and due to the termination smectite and kaolin could be preserved to the present day (Pal et al. 2000a).

7.4   Clay Minerals in Soils of the Indo-Gangetic Plains (IGP)

Srivastava et al. (1998) reported the transformation of clay minerals in a soil chrono-association comprising 5 fluvial surfaces (QGH1 to QGH5) of the IGP between Ramganga and Rapti rivers. They demonstrated that pedogenic smectite-kaolinite (Sm/K) can be considered as a potential indicator for Holocene climate changes from arid to humid conditions. The ages of QGH1 to QGH5 are <500 year BP, >500 year BP, >2500 year BP, 8000 Cal year BP and 13,500 Cal year BP, respectively. During soil formation, two major regional climatic cycles are recorded, and they are relatively arid to semi-arid cycles between 10,000–6500 year BP and 4000 year BP till present and were punctuated by a warm and humid climate. During arid climates biotite weathered to trioctahedral vermiculite and smectite in the soils that was unstable and transformed to Sm/K (please refer to

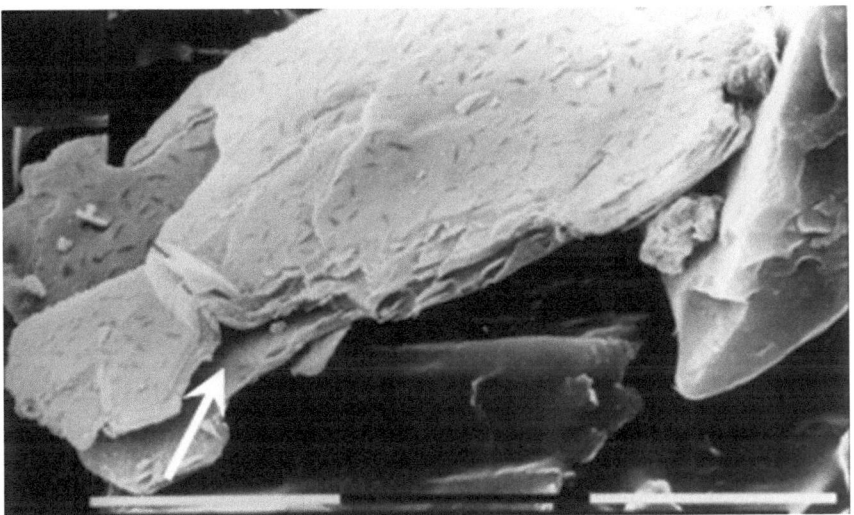

Fig. 7.1 SEM photograph of sand size biotite in soils of the IGP showing the formation of vermiculite (Adapted from Srivastava et al. 2013)

Fig. 4.3) during the following warm and humid climate phase (7400–4150 Cal year BP). When the humid climate terminated, vermiculite, smectite and Sm/K were preserved to the present. During the hot semi-arid climate that followed the humid climate, transformation of biotite into its weathering products like trioctahedral vermiculite and smectite did continue (Fig. 7.1). In arid to semi-arid climate weathering of plagioclase feldspar yielded $CaCO_3$ in soils, which is designated as pedogenic $CaCO_3$ (PC). Initiated by the formation of PC, fine clay vermiculite and smectite translocated downward in the profile as Na-clay, to make soils calcareous and sodic (Pal et al. 2003). This pedogenetic process with time becomes an example of self-terminating process (Yaalon 1971). Presence of Sm/K, trioctahedral smectite and PC in the IGP soils older than 2500 year. BP exhibit their polygenetic features (Srivastava et al. 1998).

7.5   Vertisols, Carbonate Minerals and Climate Change

In the Deccan basalt area, smectitic Vertisols occur in humid tropical (HT), sub-humid moist (SHM), sub-humid dry (SHD), semi-arid moist (SAM), semi-arid dry (SAD) and arid dry (AD) climatic environments (Pal et al. 2009). Smectites are ephemeral in HT climate as they readily transform to kaolinite (Pal et al. 1989; Bhattacharyya et al. 1993). The formation and persistence of Vertisols in HT climate has been possible because the soils are endowed with Sm/K and Ca-zeolites. These two minerals created a favourable chemical environment necessary for the formation and persistence of Vertisols in lower topographic situation

(Bhattacharyya et al. 1993, 1999). As for the formation of Vertisols huge amount clay smectite as parent material is a mandatory requirement and such amount clay smectite cannot be generated in dry climates. Therefore, it is difficult to understand the formation of Vertisols in SHM, SHD, SAM, SAD and AD climates especially when the weathering of primary minerals contributes very little towards the formation of smectites in these climatic environments (Srivastava et al. 2002; Pal et al. 2009). XRD analysis of fine clays (Fig. 7.2) indicates that smectites of Vertisols from sub-humid to arid climates are fairly well crystallized as evident from a regular series of higher order reflections and do not show any sign of transformation except for hydroxy-interlayering (HI) in the smectite interlayers (Pal et al. 2000a; Srivastava et al. 2002). Such interlayering did also occur in vermiculite of the silt and coarse clay fractions (please refer to Fig. 2.9) that further transformed to the formation of pseudo- chlorite (PCh).The hydroxy-interlayering in smectite interlayers is identified from the broadening of the low angle side of the collapsed 1. 0 nm peak of K-saturated smectite heated to 550 °C (Fig. 7.2). PCh is not a true chlorite as it shows a broad peak around 1.4 nm when K-saturated sample is heated to 550 °C (please refer to Fig. 2.9). Thus the presence of hydroxy-interlayered dioctahedral smectite (HIS), hydroxy-interlayered vermiculite (HIV, as an alteration product of biotite mica) and PCh (as a transformation product of HIV) is a unique combination of secondary minerals in size fractions of Vertisols (Pillai et al. 1996; Pacharne et al. 1996; Vaidya and Pal 2003). The hydroxy-interlayering in the vermiculite and smectite occurs when positively charged hydroxy-interlayer materials such as $[Fe_3(OH)_6]^{3+}$, $[Al_6(OH)_{15}]^{3+}$, $[Mg_2Al(OH)_6]^+$, and $[Al_3(OH)_4]^{5+}$ (Barnhisel and Bertsch 1989) enter into the inter-layer spaces at pH much below 8.3 (Jackson 1964). Moderately acidic conditions are optimal for hydroxyl interlayering of vermiculite and smectite and the optimum pH for interlayering in smectite and vermiculite is 5.0–6.0 and 4.5–5.0, respectively (Rich 1968). It is to be noted that the pH of the majority of Vertisols of sub humid to arid climates is either near to neutral or well above 8.0 throughout the profile. In mildly to moderately alkaline conditions of soils, 2:1 layer silicates suffer congruent dissolution (Pal 1985) and thus discounts the hydroxy-interlayering of smectites and vermiculites during the post depositional period of the basaltic alluvium (Pal et al. 2012a), likewise the subsequent transformation of vermiculite to PCh. The mechanism of the formation of HIS, HIV and PCh, described here, does not represent contemporary pedogenesis of Vertisols in the prevailing dry climatic conditions (Pal et al. 2012a). Vertisols of sub humid to arid climates have both NPC (relict Fe–Mn coated carbonate nodules) and PC (pedogenic $CaCO_3$) (Pal et al. 2000b, 2009). Based on ^{14}C dates of carbonate nodules, Mermut and Dasog (1986) concluded that Vertisols with Fe–Mn coated $CaCO_3$ are older soils than those with PCs that are formed in soils of dry climate (Pal et al. 2000b). NPCs were therefore, formed in a climate much wetter than the present, which ensured adequate soil water for reduction and oxidation of iron and manganese to form Fe–Mn coatings. The formation of Fe–Mn coatings indicate that Vertisols areas did experience a humid climate in the past and thus, the large amount of DSm formed in an earlier humid climate in the source area as an alteration product of plagioclase in tropical humid climates (Pal et al. 1989, 2012a;

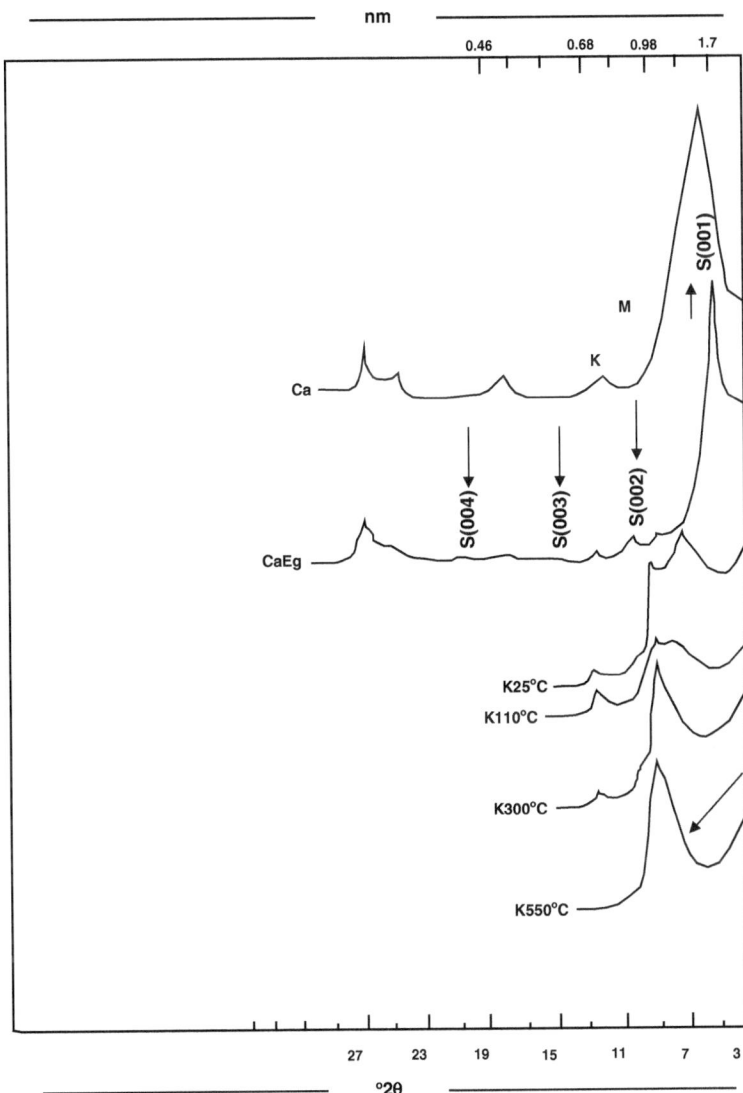

Fig. 7.2 Representative X-ray diffractograms of fairly well crystalline fine clay smectite of Vertisols despite having partial hydroxy-interlayering (Adapted from Pal 2003)

Srivastava et al. 1998), and during this weathering vermiculite transformed to HIV, which further transformed to PCh because hydroxy-interlayering in vermiculite would occur in acidic soil conditions. However, the formation of HIS in humid tropical climate did not continue as evidenced from the presence of very small amount of kaolin (KI-HIS) in the fine clay fractions. In the event of prolonged formation of HIS, the content of kaolin should have been dominant as reported by

Bhattacharyya et al. (1993) for the ferruginous Alfisols of the humid Western Ghats. The smectite of Vertisols thus formed in an earlier and more humid climate. Its crystallinity, and also the PCh were preserved in the non-leaching environment of the latter dry climates (Pal et al. 2009, 2012a, b). The ^{14}C age of soil organic carbon of Vertisols indicate soil age between 3390 to 10,187 year BP (Pal et al. 2006). This suggests that the climate did change from humid to drier climate in Peninsular India during the late Holocene (Pal et al. 2001, 2003, 2006; Deotare 2006). As a result, Vertisols in SAM, SAD and AD climates became more calcareous and sodic (Pal et al. 2006, 2009) than those of SHM and SHD climates. Therefore, they qualify to be polygenetic (Pal et al. 2001).

7.6   Implications for Climate Change from Clay Minerals Record in Drill Cores of the Ganga Plains

High resolution clay mineralogical study by Pal et al. (2012a) documents the climate change over the last 100 ka manifested in clay mineralogy of paleosols and sediments from two cores ($\sim$50 m deep) in the Ganga–Yamuna interfluve in the Himalayan Foreland Basin, India. Core sediments from the northern part of the interfluve (Indian Institute of Technology, Kanpur, IITK core) are micaceous and dominated by hydroxy-interlayered dioctahedral low-charge smectite (LCS) in fine clay fraction but by trioctahedral high-charge smectite (HCS) in silt and coarse clay fractions. In contrast, core sediments from the southern part of the interfluves (Bhognipur core) are poor in mica and both LCS and HCS are present in the upper 28 m of the core and the lower part is dominantly LCS in all size fractions. In the two cores paleosols are formed in the sub-humid to semi-arid climatic conditions and the formation of paleosols has resulted clay minerals such as 1.0–1.4 nm minerals, vermiculite, HCS and also preserved the LCS, hydroxy-interlayered vermiculite (HIV) and pseudo-chlorite (PCh), and kaolin that formed earlier in a humid climate. The preservation of LCS, HIV, kaolin and PCh bears the signature of climate shift from humid to semi-arid conditions in the Ganga Plains as their formation does not represent contemporary pedogenesis in the present alkaline chemical environment induced by the semi-arid climate. The abundance of LCS sediments in both the cores suggests the role of plagioclase weathering in the formation of LCS. The climatic records inferred from the typical clay mineral assemblages of the two interfluve cores are consistent with the Marine Isotope Stages (MIS) over the last 100 ka. Typical clay mineral assemblage in humid interglacial stages (e.g. MIS 5, 3 and 1) (Fig. 7.3) is marked by HIV, LCS and PCh formed under acidic soil conditions. In a drier climate (MIS 4 and 2), formation of trioctahedral HCS from biotite weathering and precipitation of pedogenic $CaCO_3$ were the dominant processes (Fig. 7.3). The formation of pedogenic $CaCO_3$ created conducive environment for illuviation of clays forming argillic (Bt) horizon in the paleosols of the interfluve (Srivastava et al. 2010).

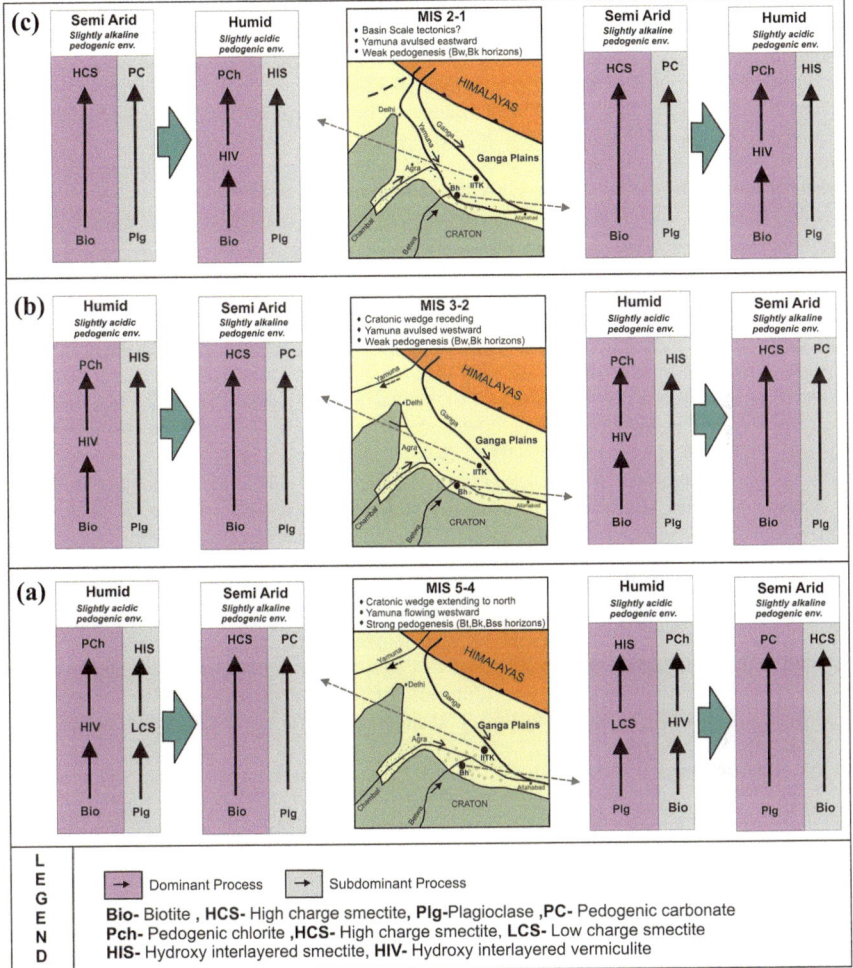

Fig. 7.3 A schematic climate-driven model for clay mineral alteration for the last 100 ka in northern and southern parts of the Ganga–Yamuna interfluve represented by IITK and Bhognipur cores (Adapted from Pal et al. 2012a)

References

Barnhisel RI, Bertsch PM (1989) Chlorites and hydroxy-interlayered vermiculites and smectite. In: Dixon JB, Weed SB (eds) Minerals in soil environments. Soil Science Society of America Book Series (Number 1), Second edition. Wisconsin, USA, pp 729–788

Beckmann GG (1984) Paleosols, pedoderms, and problems in presenting pedological data. Aust Geogr 16:15–21

Bhattacharyya T, Pal DK, Deshpande SB (1993) Genesis and transformation of minerals in the formation of red (Alfisols) and black (Inceptisols and Vertisols) soils on Deccan Basalt in the Western Ghats, India. J Soil Sci 44:159–171

Bhattacharyya T, Pal DK, Srivastava P (1999) Role of zeolites in persistence of high altitude ferruginous Alfisols of the humid tropical Western Ghats, India. Geoderma 90:263–276

Chandran P, Ray SK, Bhattacharyya T, Krishnan P, Pal DK (2000) Clay minerals in two ferruginous soils of southern India. Clay Res 19:77–85

Deotare BC (2006) Late Holocene climatic change: archaeological evidence from the Purna Basin, Maharashtra. J Geol Soc India 68:517–526

Fenwick I (1985) Paleosols: problems of recognition and interpretation. In: Boardman J (ed) Soils and quaternary landscape evolution. Willey, Chichester, pp 3–21

Jackson ML (1964) Chemical composition of soils. In: Bear FE (ed) Chemistry of the soil. Oxford and IBH Publishing Co., Calcutta, pp 71–141

Jenkins DA (1985) Chemical and mineralogical composition in the identification of paleosols. In: Boardman J (ed) Soils and quaternary landscape evolution. Wiley, New York, pp 23–43

Keller WD (1970) Environmental aspects of clay minerals. J Sediment Petrol 40:788–813

Mermut AR, Dasog GS (1986) Nature and micromorphology of carbonate glaebules in some Vertisols of India. Soil Sci Soc Am J 50:382–391

Murali V, Krishnamurti GSR, Sarma VAK (1978) Clay mineral distribution in two toposequences of tropical soils of India. Geoderma 20:257–269

Pacharne T, Pal DK, Deshpande SB (1996) Genesis and transformation of clay minerals in the formation of ferruginous (Inceptisols) and black (Vertisols) soils in the Saptadhara Watershed of Nagpur district, Maharashtra. J Indian Soc Soil Sci 44(300):309

Pal DK (1985) Potassium release from muscovite and biotite under alkaline conditions. Pedologie (Ghent) 35:133–146

Pal DK (2003) Significance of clays, clay and other minerals in the formation and management of Indian soils. J Indian Soc Soil Sci 51:338–364

Pal DK, Deshpande SB (1987) Genesis of clay minerals in a red and black complex soil of southern India. Clay Res 6:6–13

Pal DK, Deshpande SB, Venugopal KR, Kalbande AR (1989) Formation of di- and trioctahedral smectite as an evidence for paleoclimatic changes in southern and central Peninsular India. Geoderma 45:175–184

Pal DK, Bhattacharyya T, Deshpande SB, Sarma VAK, Velayutham M (2000a) Significance of minerals in soil environment of India, NBSS review series 1. NBSS&LUP, Nagpur 68 pp

Pal DK, Dasog GS, Vadivelu S, Ahuja RL, Bhattacharyya T (2000b) Secondary calcium carbonate in soils of arid and semi-arid regions of India. In: Lal R, Kimble JM, Eswaran H, Stewart BA (eds) Global climate change and pedogenic carbonates. Lewis Publishers, Boca Raton, pp 149–185

Pal DK, Balpande SS, Srivastava P (2001) Polygenetic Vertisols of the Purna Valley of Central India. Catena 43:231–249

Pal DK, Srivastava P, Bhattacharyya T (2003) Clay illuviation in calcareous soils of the semi-arid part of the Indo-Gangetic Plains, India. Geoderma 115:177–192

Pal DK, Bhattacharyya T, Ray SK, Chandran P, Srivastava P, Durge SL, Bhuse SR (2006) Significance of soil modifiers (Ca-zeolites and gypsum) in naturally degraded Vertisols of the Peninsular India in redefining the sodic soils. Geoderma 136:210–228

Pal DK, Bhattacharyya T, Chandran P, Ray SK, Satyavathi PLA, Durge SL, Raja P, Maurya UK (2009) Vertisols (cracking clay soils) in a climosequence of Peninsular India: evidence for Holocene climate changes. Quatern Int 209:6–21

Pal DK, Bhattacharyya T, Sinha R, Srivastava P, Dasgupta AS, Chandran P, Ray SK, Nimje A (2012a) Clay minerals record from late quaternary drill cores of the Ganga Plains and their implications for provenance and climate change in the Himalayan Foreland. Palaeogeogr Palaeoclimatol Palaeoecol 356–357:27–37

Pal DK, Wani SP, Sahrawat KL (2012b) Vertisols of tropical Indian environments: pedology and edaphology. Geoderma 189–190:28–49

Pillai M, Pal DK, Deshpande SB (1996) Distribution of clay minerals and their genesis in ferruginous and black soils occurring in close proximity on Deccan basalt plateau of Nagpur district, Maharashtra. J Indian Soc Soil Sci 44:500–507

Rengasamy P, Sarma VAK, Murthy RS, Krishnamurti GSR (1978) Mineralogy, genesis and classification of ferruginous soils of the eastern Mysore plateau, India. J Soil Sci 29:431–445

Rich CI (1968) Hydroxy-interlayering in expansible layer silicates. Clays Clay Miner 16:15–30

Singer A (1980) The paleoclimatic interpretation of clay minerals in soils and weathering profiles. Earth-Sci Rev 15:303–326

Srivastava P, Parkash B, Pal DK (1998) Clay minerals in soils as evidence of Holocene climatic change, central Indo-Gangetic Plains, north-central India. Quatern Res 50:230–239

Srivastava P, Bhattacharyya T, Pal DK (2002) Significance of the formation of calcium carbonate minerals in the pedogenesis and management of cracking clay soils (Vertisols) of India. Clays Clay Miner 50:111–126

Srivastava P, Singh AK, Parkash B, Singh AK, Rajak M (2007) Paleoclimatic implications of micromorphic features of quaternary paleosols of NW Himalayas and polygenetic soils of the Gangetic Plains—a comparative study. Catena 70:169–184

Srivastava P, Rajak M, Singh LP (2009) Late quaternary alluvial fans and paleosols of the Kangra Basin, NW Himalaya: Tectonic and paleoclimatic implications. Catena 76:135–154

Srivastava P, Rajak MK, Sinha R, Pal DK, Bhattacharyya T (2010) A high resolution micromorphological record of the late quaternary paleosols from Ganga-Yamuna Interfluve: stratigraphic and paleoclimatic implications. Quatern Int 227:127–142

Srivastava P, Pal DK, Bhattacharyya T (2013) Mineral formation in soils and sediments as signatures of climate change. In: Bhattacharyya T, Pal DK, Sarkar D, Wani SP (eds) Climate change and agriculture. Studium Press, New Delhi, pp 223–234

Tardy Y, Bocquier G, Paquet H, Millot G (1973) Formation of clay from granite and its distribution in relation to climate and topography. Geoderma 10:271–284

Vaidya PH, Pal DK (2003) Mineralogy of Vertisols of the Pedhi Watershed of Maharashtra. Clay Res 22:43–58

Valentine KWG, Dalrymple JB (1976) Quaternary buried paleosols: a critical review. Quatern Res 6:209–220

Wright VP (1986) Paleosols their recognition and interpretation. Blackwell Scientific Publications, Oxford

Yaalon DH (1971) Soil forming processes in time and space. In: Yaalon DH (ed) Paleopedology. Israel University Press, Jerusalem

Chapter 8
Linking Minerals to Selected Soil Bulk Properties

Abstract Minerals in soils are the result of both inheritance and authigenic formation of both primary and secondary minerals. Comprehensive reviews on the soil clay minerals and other minerals in the past indicate that there are not many attempts to show the influence of minerals in soil genesis and management. A search for links between mineralogy and soil properties of agricultural importance is likely to be difficult because many a time the description of minerals actually present in a soil is inadequate or incomplete. Further, as soil minerals often differ from "type" minerals, it is very much necessary to investigate the properties of these minerals relevant to the properties of the soil in bulk. Despite our general understanding on the role of minerals in soils, it is necessary to investigate the properties of the minerals, especially clay minerals, their mixtures and surface modifications in the form that they occur in the soil. From the few examples under different agro-climatic situations cited it is evident that unless the mineralogical description is accurate enough for the purpose intended, it would not be prudent to look for their significance in soils. With the use of high resolution mineralogy, identification and explanation of many enigmatic situations in soils can be conveniently solved. Therefore, the advanced information developed provides adequate mineralogical database that would explain discretely many unresolved issues of the nutrient management in terms of specific soil minerals in general, and clay minerals in particular and their significance in soil as a sustainable medium for plant growth. It is hoped that this synthesis would help to assess the health and quality of soils while developing suitable management practices to enhance and sustain their productivity in the 21st century.

Keywords Soil minerals · Characterisation of clay minerals · High end instruments · Nutrient management · Suitable management practices

© Springer International Publishing AG 2017   127
D.K. Pal, *A Treatise of Indian and Tropical Soils*,
DOI 10.1007/978-3-319-49439-5_8

8.1   Introduction

Minerals in soils are the result of both inheritance and authigenic formation of both primary and secondary minerals. Comprehensive reviews on the soil clay minerals and other minerals in the past indicate that there are not many attempts to show the influence of minerals in soil genesis and management (Pal et al. 2012a). A review in the past on this subject and related issues (Newman 1984; Ghosh 1997; Wilson 1999) pointed out that the increase in agricultural productivity during the middle of the twentieth century was due to human influence, which vastly improved the soil fertility on all types of soil. However, actual productivity does vary widely and this fact shows that there are other factors of productivity beyond our control. One intrinsic factor in yield variation is soil type, which is related to the soil composition and its position in the landscape. Clay is an important soil constituent controlling its properties. Despite the fact that there is ample evidence to show that the amount of clay in a soil has a very important bearing on the genesis, characteristics, and physical and chemical properties of soils, it would be more appropriate to see what significance clay mineral *type* and other soil minerals have in soils. A search for links between mineralogy and soil properties of agricultural importance is likely to be difficult because many a time the description of minerals actually present in a soil is inadequate or incomplete. Further, as soil minerals often differ from "type" minerals, it is very much necessary to investigate the properties of these minerals relevant to the properties of the soil in bulk. In this endeavour, Pal et al. (2000a, 2012a) demonstrated a good number of examples that indicated despite soil clay minerals being a mixture of several components, adequate description is possible. Synthesis of the present dataset on the nature and characteristics of primary and secondary minerals of Indian soils has established a link between minerals and selected bulk soil properties. It is hoped that this synthesis would help to assess the health and quality of soils while developing suitable management practices to enhance and sustain their productivity in the 21st century.

8.2   Clay and Other Minerals in Adsorption and Desorption of Major Nutrients

8.2.1   Organic and Inorganic Carbon

8.2.1.1   Vertisols

Vertisols occupy only 8.1% of the total geographical area of the country but have 29% share in total soil organic carbon (SOC) stock (0–150 cm soil depth) of 29.32 Pg, indicating the role of surface area (due to the dominating amount of smectite clay minerals) in accumulating OC. Vertisols and soils with vertic characters under agricultural land uses under both short and long-long term experiments also displayed their potential to sequester OC under both arable and submerged conditions and they are also capable enough to sequester OC even in humid climates (Pal et al. 2015). Results on Vertisols indicated that legume-based improved management (IM) advocated by ICRISAT (detailed in Chap. 6) could sequester OC at the rate of 5 mg year^{-1} in the first 100 cm soil depth even without FYM and gypsum addition. When FYM added (10 Mg FYM ha^{-1}) along with 100% of recommended doses of NPK, Vertisols showed potential to sequester an additional amount of 330 kg OC ha^{-1} year^{-1} (Pal et al. 2015).

Zeolites improve hydraulic properties of soils by enriching soil exchange complex with Ca^{2+} ions in arid to HT climate (Pal et al. 2013). In order to follow the decomposition resistance of soil organic matter under high-temperature, experiments with Ca-zeolite and organic manure showed a slight increase in C/N ratio in soils of the Philippines, Paraguay and Japan. In addition, carbon accumulation in humic fractions as well as the degree of humification and aromaticity of humic acids increased (Truc and Yoshida 2011). In semi-arid dry region of India, zeolitic (heulandite) Vertisols (Teligi soils, Bellary, Karnataka; Jhalipura soils, Kota, Rajasthan; Jajapur, Mehboobnagar, Andhra Pradesh) (Pal et al. 2003), under wetland rice–rice/rice–wheat system showed wider C/N ratio (Table 8.1), indicating enough potential to sequester atmospheric carbon (Sahrawat et al. 2005). This suggests that the presence of zeolites could be beneficial for soil organic matter conservation under global warming (Pal et al. 2013).

8.2.1.2 Red Ferruginous (RF) Soils

Total stock of soil organic carbon (SOC) is 9.55 Pg in the first 30 cm depth, and out of this stock, Indian soils of the arid (cold arid and hot arid) and semiarid climate have a share of about 40%, the soils of sub-humid, humid and per-humid 47%, and the Ultisols of about 1% (Bhattacharyya et al. 2000a). Therefore, RF soils of HT climate are not impoverished in OC. RF soils under dry climate contain <1% OC but SAT Alfisols under permanent fallow with grass cover could sequester OC up

Table 8.1 Selected properties in surface (0–30 cm) soil samples of zeolitic Vertisols under rice cultivation

Benchmark soil series	District/state	Soil taxonomy	pH (1:2)	Clay CEC	SOC (%)	Total N (%)	SOC: N
Jhalipura	Kota/Rajasthan	Typic Haplusterts	8.1	77	0.53	0.0443	12:1
Jajapur 1	Mehboobnagar/Andhra Pradesh	Sodic Haplusterts	8.5	62	0.88	0.082	11:1
Teligi	Bellary/Karnataka	Sodic Haplusterts	8.0	90	1.03	0.062	17:1
Teligi1	Bellary/Karnataka	Sodic Haplusterts	7.8	99	0.88	0.0551	14:1

Adapted from Sahrawat et al. (2005) and Pal et al. (2003)

to ~ 1.5, $\sim 1\%$ under horticulture and 1.78% under forest in the first 30 cm of the profile (Bhattacharyya et al. 2014; Pal et al. 2015). A short term experiment on SAT Alfisols under sorghum-castor bean rotation (Sharma et al. 2005) indicated that conventional tillage with the application of Glyricidia loppings along with 90 kg ha^{-1} N provided the best soil quality index (SQI), increased vegetative growth and root biomass, which in turn enhanced soil organic matter. Srinivasarao et al. (2013) pointed out that the importance of identification and management factors that cause enhanced C sequestration in SAT soils. They also indicated that recommended management practices for SAT soils that include locally available organic resources, are capable of improving the SOC sequestration.

Ultisols, Alfisols and Mollisols of HT climate have OC concentration ranges from 1.0 to 5% (please refer to Table 3.2), which is relatively high as compared to SAT Alfisols. These soils are developed under thermic, hyperthermic (Nilgiri Hills in Kerala and Tamil Nadu, Manipur, Meghalaya, Nagaland, Arunachal Pradesh, Assam, Tripura and Mizoram), isohyperthermic (Andaman and Nicobar, Kerala, Tamil Nadu, Madhya Pradesh and Maharashtra) soil temperature regime, and udic (Andaman and Nicobar, Arunachal Pradesh, Manipur, Meghalaya, Assam, Tripura, Nagaland, Mizoram, Nilgiri Hills in Kerala and Tamil Nadu) and ustic (Kerala, Karnataka, Tamil Nadu, Madhya Pradesh and Maharashtra) soil moisture regime (Bhattacharyya et al. 2009). Soils in a wet climate under forest have high OC content, sufficient to qualify as Mollisols. The OC addition to Ultisols and Alfisols has been possible as a result of, favourable soil temperature and moisture regime. It is known that 2:1 expanding clay minerals provide higher surface area for OC accumulation. But a high positive OC balance in kaolin dominated RF soils clearly indicates a positive role of this mineral and other hydroxy-interlayered clay minerals because kaolin (like the other hydroxy-interlayered minerals) often shows a relatively high value of CEC ~ 30 cmol (p+) kg^{-1} (Ray et al. 2001) than that of well crystalline kaolinite. Therefore, besides the dominating effect of humid climate in cooler winter months with profuse vegetation, the soil substrate quality in terms of larger surface area is of fundamental importance in OC sequestration in soils (Bhattacharyya et al. 2014; Pal et al. 2015).

The formation and persistence of acidic and fairly weathered Mollisols and OC rich (>1%) Alfisols on zeolitic Deccan basalt of HT climate (Bhattacharyya et al. 2005, 2006) in contrast to commonly found alkaline Mollisols in temperate humid climate and Ultisols of HT climate, suggests that Ca-zeolite is an another important substrate for OC sequestration. The Ca-zeolites (as soil modifier) are sources of bases to prevent complete transformation of smectite to kaolinite by maintaining relatively high base saturation level in acidic Mollisols and Alfisols.

8.2.1.3 IGP Soils

The sustainability ratings of some soil series of the IGP for the rice-wheat cropping system indicate many soil constraints, including low soil organic carbon (SOC) (Bhattacharyya et al. 2004). The SOC and inorganic carbon (SIC) stocks of the IGP (0–150 cm depth) are 2.0 and 4.58 Pg, respectively. The SOC stock of the IGP constitutes 6.45% of the total SOC stock of India, 0.30% of the tropical regions and 0.09% of the world in the first 30 cm depth of the soil. The corresponding values of SIC are 3.20% for India, 0.17% for the tropical regions and 0.06% for the world, respectively. Thus the soils of the IGP are impoverished in OC compared to tropical regions and the world in general and to India in particular (Bhattacharyya et al. 2004). For soils of the IGP, Bhattacharyya and co-workers (2004) indicated that five AESRs (warm to hot moist, hot moist humid to per-humid, warm to hot per-humid, warm to hot per-humid and warm to hot per-humid) covering 6% area of the IGP are in the sufficient zone of OC and the remaining nine AESRs covering 94% area are under deficient zone. However, some areas under humid climate do not have sufficient SOC because of intensive agricultural practices (Abrol and Gupta 1998).

It is generally observed that among the agricultural systems such as rice–wheat, rice–rice, cotton-wheat and groundnut, the cereal-based systems contribute to higher accumulation and stabilization of organic matter, especially in rice–wheat systems (Bhattacharyya et al. 2007a, b), because decomposition of organic matter in the absence of oxygen is slow, incomplete and inefficient. It is also due to the formation of recalcitrant complexes with organic matter in these soils that render organic matter less prone to microbial attack (Sahrawat 2004a, b). Thus the SOC built up under submerged conditions is the reason for high status of organic matter and productivity in rice–rice and rice–wheat systems in the IGP. The Indian Council of Agricultural Research-National Bureau of Soil Survey and Land Use Planning (ICAR-NBSS & LUP) reassessed the SOC stock of the IGP in 2005, and observed that there has been 30% increase in SOC in 0–20 cm depth over the value assessed in 1980. Soils of the sub-humid part of the IGP (Vertic Endoaqualfs at Mohanpur, West Bengal, Pal et al. 2010) with double or triple rice-based agricultural land uses under long-term experiments showed their potentiality to sequester OC under submerged conditions and still show potential to sequester OC (Majumder et al. 2007; Mandal et al. 2007, 2008), suggesting the benefit of higher soil surface area (caused by 2:1 expanding clay mineral) in better OC sequestration even in sub-humid to humid climates.

The SIC stocks in soils of the arid and semi-arid ecosystems of the IGP seem to be useful during the establishment of vegetation by appropriate ameliorative methods in these soils, as the plant roots can dissolve the immobile $CaCO_3$ and can ultimately trigger the process of Ca^{2+} release in the soil and thus act as a natural ameliorant for sodic soils (Bhattacharyya et al. 2004). The benefit of such chemical reaction is realized recently when the naturally occurring grassland systems was continued on IGP sodic soils for almost four decades (Jangra et al. 2015). These authors report that by increasing plant biomass a marked improvement in soil organic carbon was observed (from 0.20 to 0.44%). The extensive root system of *Desmostachya bipinnata*, the dominant grass, through biological production of carbonic acid by the roots, seems to play an important role in solubilisation of native $CaCO_3$ present in the sodic soils. The dissolved Ca^{2+} ions caused a substantial decrease in soil pH, electrical conductivity and an increase in exchangeable

Ca and Mg. The huge SIC stock thus remains as a hidden treasure that would improve the drainage and help in the establishment of vegetation and also sequestering OC in the soils (Bhattacharyya et al. 2004). The overall increase in SOC stock in the benchmark spots under agriculture, practised for the last 25 years, suggests that agricultural management practices of the National Agricultural Research System (NARS) did not cause any decline in SOC (Bhattacharyya et al. 2007a, b).

8.2.2 Nitrogen

8.2.2.1 Vertisols

Increasing demand for nitrogen (N) fertilizers to produce food grains has always stimulated research to gain knowledge on the various forms of N in soils. One of the forms of mineral N is fixed NH_4-N, and several reports indicate that many tropical Vertisols are endowed with large amounts of fixed ammonium (Sahrawat 1995). Vermiculites, illites and smectites are often considered able to fix NH_4-N (Nommik and Vahtras 1982). Smectites have no selectivity for non-hydrated monovalent cations such as K^+ because of their low layer charge (Brindley 1966). NH_4^+ ion, also a non-hydrated monovalent cation with almost the same ionic radius as K, is not expected to be fixed in the interlayers of smectites. It is equally difficult to understand the NH_4 ion fixing capacity of illites, because they do not expand on being saturated with divalent cations (Sarma 1976). Earlier reports indicate that Vertisols developed in the basaltic alluvium of the Deccan basalt of Peninsular India, do not contain vermiculite (Dhillon and Dhillon 1991). However, a recent report indicates that vermiculite content in such soils ranges from 2.0 to 3.5% in the silt, 3.5 to 10% in the coarse clay and 5.0 to 9.5% in the fine clay fractions (Pal and Durge 1987). The identification of vermiculite by XRD analysis in different soil size fractions is fraught with some difficulty in the ubiquitous presence of chlorite. Its presence is resolved by following the progressive reinforcement of the 1.0 nm peak of mica while heating the K-saturated samples at 25, 110, 300 and 550 °C, and its quantity is estimated semi-quantitatively (Pal et al. 2012b). Thus, it would be prudent to attribute the observed NH_4-N fixation in Vertisols (Sahrawat 1995) to the presence of vermiculite only.

Zeolites are known to have pronounced selectivity for NH_4^+-N over $Ca^{2+,}$ Mg^{2+} and Na^+; and it is difficult to remove NH_4^+ from zeolite exchange sites by these less selective cations. Therefore, NH_4^+ is slowly released, however its rate of release from vermiculite and zeolite in a zeolitic Vertisols is not yet known. Thus, it would not be prudent to attribute the observed NH_4-N fixation in Vertisols (Sahrawat 1995) entirely to the presence of vermiculite only (Pal et al. 2012b). Such a basic understanding is essential to include fixed NH_4-N in assessing the potential of N available in tropical soils in general and the Vertisols in particular A new research initiative in this direction is thus awaited.

Zeolites have the ability to protect $NH_4{}^+$ on zeolite exchange sites from microbial conversion of $NH_4{}^+$ to $NO_3{}^-$ because nitrifying bacteria are too large to fit into the channels and cages within zeolite structure where $NH_4{}^+$ ion resides on exchange sites (Ming and Allen 2001). This way the protection of $NH_4{}^+$ suggests that emission of N_2O from organic (farmyard manure) and inorganic N fertilizers would amount to a small fraction of the total world greenhouse gas emissions from zeolitic soils, because out of 500,000 km^2 Deccan basalt area in the Indian subcontinent (Pal et al. 2000a), zeolitic soil is expected to cover a considerable part (Pal et al. 2013). Bhattacharyya et al. (2015) in their recent study have updated the knowledge on the occurrence of Ca-rich zeolites in cracking clay soils in the IGP area and black soil region (BSR), and reported an area ~ 2.8 m ha, of which BSR and IGP occupy ~ 92 and $\sim 8\%$, respectively. In addition, this study provides an approximate map on the distribution of Ca-rich zeolites in Indian soils. Such map would be of much help to include fixed NH_4-N in assessing the N available in zeolitic soils, especially when N_2O emission from Indian agricultural soils is a small fraction (about 1%) of the global warming caused by CO_2 emissions (Bhatia et al. 2004, 2012).

8.2.2.2 RF Soils

RF soils need N fertilization for agricultural production. One of the forms of mineral N in soils is fixed NH_4-N and some selected tropical soils contain large amounts of fixed N (Dalal 1977). Information on fixed N in soils of HT climate, especially under paddy or lowland rice is scarce, despite the fact that Fe rich soils under submerged environments do sequester good amount of OC (Sahrawat 2004a). Iron (Fe) is present in large amounts in RF soils and reducible Fe influence the NH_4 production or N mineralization in submerged soils. In the absence of oxygen, ferric Fe serves as an electron acceptor and affects organic matter oxidation and NH_4 production (Sahrawat 2004b). Similar kind of pedochemical reactions in OC rich RF soils of HT climate under rice in Kerala, Goa, and NEH, is expected to benefit the soils in enriching their fixed NH_4 status and minimizing the effect of N fertilizers. Information on fixed NH_4-N is also scarce for RF soils of the semi-arid tropics (SAT) (Burford and Sahrawat 1989). Realizing its importance in the N economy of RF soils, Sahrawat (1995) determined the NH_4-N distribution in a benchmark Alfisol (Patancheru soils at ICRISAT, Patancheru farm, Telangana). The amount of fixed NH_4-N as per cent of total N varied from 14.0 to 30.8% and the values generally increased with soil depth. Vermiculites are the only clay minerals to fix NH_4-N. Patancheru soils do contain vermiculites in their silt, coarse and fine clay fractions, and the translocation of fine clay-vermiculite has enriched the subsoils with vermiculite (Pal et al. 2012b) that explains the observed increasing fixation of NH_4-N with soil depth (Sahrawat 1995). In soils of HT climate, NH_4-N is also expected to get fixed in discreet vermiculite and also in its hydroxy-interlayered counterpart. The $NH_4{}^+$ ion, also a non-hydrated monovalent cation with almost the same ionic radius as K, is expected to be fixed in the vermiculite interlayers, and this way NH_4^+ ion may be protected from microbial conversion of NH_4^+ to NO_3^-. The prevention of such conversion suggests that the emission of N_2O from organic (farm yard manure) and inorganic N fertilizers from RF soils under arable and submerged conditions would also amount to a small fraction of the total world greenhouse gas emission (Bhatia et al. 2004, 2012). Like in Vertisols such basic understanding is essential to include fixed NH_4-N for assessing potentiality of available N in RF soils.

8.2.2.3 IGP Soils

Like any other important soil types of India, the IGP soils also need N fertilizers to sustain crop productivity. The amount of fixed N in the IGP soils is very rarely reported. Therefore, it is difficult to highlight the role of specific clay minerals capable to fix NH_4^+ ions. However, it will not be too imprudent to anticipate the role of 2:1 clay minerals in fixing NH_4^+ ions in view of the observed K adsorption and fixation reactions (Pal et al. 2000a). In general, micaceous IGP soils do contain vermiculite and low charge vermiculite or high charge smectite in the finer soil fractions (Pal et al. 2000a). In view of their K adsorption capacity, it can be safely assumed that the IGP soils can also adsorb NH_4^+ ions and get fixed in the interlayer of vermiculite and smectite which are in considerable amount (Pal et al. 2000a). This way NH_4^+ may be protected from microbial conversion of NH_4^+ to NO_3^-. The prevention of such conversion would minimize the emission of N_2O from organic and inorganic N fertilizers from the IGP soils under arable and submerged conditions. However, fresh research initiative in this area of soil research will highlight the extent of fixed N in the IGP soils of different bio-climates.

8.2.3   Phosphorus

8.2.3.1 Vertisols

A critical analysis of the findings of several researchers on phosphorus (P) adsorption by soil minerals (Sanyal and DeDatta 1991) indicated a significant correlation of P sorption parameters with clay content, and these authors proposed that this is a mere reflection of the effect of specific surface area on P adsorption. In soils, hydrous oxides of iron and aluminium occur as fine coatings on the surfaces of clay minerals (Haynes 1983), and these coatings have large specific surface areas that can adsorb large amounts of added P. This characteristic suggests that crystalline aluminosilicate minerals merely play a secondary role in P adsorption (Ryden and Pratt 1980). Fine clay smectites of Vertisols of HT, SHM, SHD, SAM, SAD, and AD of Peninsular India are partially hydroxy-interlayered (Pal et al. 2009). The hydroxy interlayering in smectite interlayers is not a contemporary pedogenic process because in the prevailing mild to moderately alkaline pH conditions, the hydroxides of iron and aluminium cannot remain as positively charged cations to enter the negative environment of the interlayers of smectites (Pal et al. 2012c). The presence of hydroxy-interlayered smectite (HIS) in the fine clay fractions indicates that the hydroxy-interlayering in the smectite interlayers did occur when positively charged hydroxy interlayer materials entered into the interlayer spaces at a pH far below 8.3 (Pal et al. 2012c). Moderately acidic conditions are optimal for the hydroxy-Al interlayering of smectite, and the optimum pH for interlayering in smectite is 5.0–6.0 (Rich 1968a, b). The pH of the Vertisols is mildly to moderately alkaline, which would favour congruent dissolution of 2:1 layer silicates (Pal 1985). This scenario discounts the hydroxy-interlayering of smectites during the formation of Vertisols in the Holocene period (Pal et al. 2012c) and the creation of any positively charged hydroxides that can fix added P, as in highly weathered acidic soils. Therefore, the highest surface area of smectite and/or hydroxides of iron and aluminium with no positive sites play a small role in the adsorption of added negatively charged phosphate ions in Vertisols. This supports the ICRISAT's classical experimental observations on P adsorption and desorption on Vertisols (Kasireddipalli soils), which clearly indicate that P adsorption is not a major problem in the Vertisols and that all the adsorbed P is easily exchangeable by P^{32} and a small amount of P is adsorbed in the non-exchangeable form (Sahrawat and Warren 1989; Shailaja and Sahrawat 1994; Warren and Sahrawat 1993). ICRISAT (1988) envisaged that $CaCO_3$ could adsorb P because the effective sorption by $CaCO_3$ is not well understood, and P adsorption is not always related to $CaCO_3$ content but to the quality of the $CaCO_3$. The SAT Vertisols contain $CaCO_3$ of both nonpedogenic (NPC) and pedogenic (PC) origin, and both of them effervesce with HCl and cannot be distinguished without examining the soil thin sections under a microscope (Pal et al. 2000a). During the formation of Vertisols in the SAT environment, NPCs (pedorelict) dissolve, and the soluble Ca^{2+} ions released from NPCs become precipitated as PC at a pH of approximately 8.2 and may also react with phosphate ions to form Ca-P. Both PC and Ca-P may have the least solubility in the prevailing mild to moderately alkaline pH conditions of Vertisols. This makes it clear why Ca-P in the SAT Vertisols has a dominant share among the other soil P compounds, such as Fe–P and Al–P. Thus less soluble Ca-P causes a very low level of soluble extractable P (<5 mg kg^{-1} soil) by Olsen's method (ICRISAT 1988). It is interesting to note that grain sorghum grown on Vertisols responds little to applied P unless the level of Olsen's P was <2.5 mg kg^{-1} soil (ICRISAT 1988). Additionally, some leguminous crops such as chickpea and pigeon pea do not respond well to fertilizer P than sorghum and pearl millet (ICRISAT 1981) because the root systems of chickpea exude organic acids (malic or citric) (ICRISAT 1988) and those of pigeon pea produce piscidic acid (Ae et al. 1990). These acids dissolve Ca–P and Fe–P, making more P available to the plants. The root exudates containing such organic acids and the rootlets in the soil through which rainwater passes, or other sources of CO_2, can cause an increase in the solubility of PC and Ca–P. The improved management (including pigeon pea) in the long-term heritage watershed experiment at the ICRISAT Center, Patancheru, under rain-fed conditions (Pal et al. 2012a, b) indicates that during the last 24 years, the rate of dissolution of $CaCO_3$ was 21 mg year^{-1} in the first 100 cm of the Kasireddipalli soils, which caused a slight increase in exchangeable Ca/Mg and a decrease in pH (Pal et al. 2012a, b). The rate of dissolution of Ca–P under the present improved management system is sufficient, as it does not warrant the application of a high dose of added P fertilizer. However, predicting a time scale when soils will be devoid of Ca–P is difficult unless a new research initiative in this direction is taken up.

Many SAT Vertisols contain zeolites (Pal et al. 2006a) and thus attention is needed to follow the role of zeolites on P adsorption and desorption phenomenon. Studies indicate that the dissolution of apatite-rich phosphate rock is enhanced by the exchange of dissolved Ca^{2+} on to zeolite exchange sites. The addition of NH_4^+-, H^+-, or Na^+-exchanged, clinoptilolite-rich tuff significantly increased solution P concentration when compared with phosphate rock without zeolite additions (Ming and Allen 2001). Soils occurring in the Deccan basalt areas under semiarid and HT climate, contain heulandite [(Na, K)Ca$_4$(Al$_9$Si$_{27}$O$_{72}$) 24H$_2$O], which is rich in Ca^{2+} ions, and thus the soils are highly base-saturated. Therefore, the scope of P fertilization by dissolution and ion exchange with zeolite in such soils is expected to be limited. In smectitic Vertisols of the Deccan basalt areas, P adsorption is not a major problem and all the adsorbed P is easily exchangeable by P^{32} and a small amount is adsorbed in the nonexchangeable form (Sahrawat and Warren 1989; Warren and Sahrawat 1993).

8.2.3.2 RF Soils

Large doses of P are required for moderate to highly acidic RF soils to get desired crop response even when crop requirement of P is relatively low (Datta 2013). Nature and amount of clay, organic matter, and hydrous oxides of iron and aluminium affect P adsorption in such soils (Sanyal and DeDatta 1991).

It is often reported that soils rich in clay kaolinite may contribute to P adsorption in highly weathered soils of HT climate. But the added P is adsorbed more in soils with low pH wherein activity of iron and aluminium is high. Thus, the Fe and Al-oxy-hydroxides ordinarily present in RF soils have phosphate fixing ability. These hydroxides occur as fine coatings on surfaces of silicate clay minerals (Haynes 1983) and have large specific surface area that causes adsorption of large amounts of added P. This physical adsorption implies that in P adsorption these oxy-hydroxides have a merely secondary role (Ryden and Pratt 1980). But, these minerals can adsorb negatively charged phosphate ions only when they remain as cations in highly acidic medium. In many Alfisols and Ultisols of India are highly acidic and their KCl pH values remain close to equal or greater than water pH (Bhattacharyya et al. 2000b; Chandran et al. 2004, 2005), indicating the presence of gibbsite and/or poorly crystalline materials (Smith 1986). A negative/zero/positive ΔpH indicates the presence of variable charge minerals such as gibbsite and/or sesquioxides (Bhattacharyya et al. 1994). Therefore, gibbsite and/or sesquioxides showing a positive ΔpH could be a better substrate to absorb negatively charged phosphate ions, and this is evident from the continuous increment in yield of rice up to 120 kg P_2O_5 ha^{-1} in gibbsitic soils of Meghalaya (Datta 2013). This suggests that the highest surface area of 2:1 expanding silicate clay minerals and/or Fe and Al-oxy-hydroxides with no positive sites, have little role in the adsorption of added negatively charged phosphate ions in mild to moderately acidic soils.

Adsorption of P is observed in mildly acidic SAT Alfisols (Patancheru soils), which contain hydroxy interlayered vermiculite (HIV) in the silt and coarse clay fractions, and hydroxy-interlayered dioctahedral smectite (HIS) in the coarse and fine clays (Pal and Deshpande 1987). HIV and HIS minerals may not be responsible for P adsorption because the hydroxy-interlayering of vermiculite and smectite is not a part of the contemporary pedogenic process in the prevailing mild acidic pH conditions. Under such mild acidic pH, the Fe and Al-oxy-hydroxides do not exist as positively charged cations and thus cannot enter the negative environment of the interlayers of vermiculites and smectites (Pal et al. 2012b). Therefore, the formation of HIV and HIS did occur when positively charged hydroxy interlayer materials enter into the interlayer space at a pH 5.0–6.0 (Rich 1968a). The observed P adsorption is thus related to the formation of Ca–P as Patancheru soils like any other SAT Alfisols, contain $\sim 1\%$ $CaCO_3$ (Pal et al. 2014). Calcification is one of the contemporary pedogenetic processes in soils of SAT environments (Pal et al. 2000b).

8.2.3.3 IGP Soils

Role of Fe and Al-oxy-hydroxides in fixing the added P is now well understood. The IGP soils with neutral to alkaline pH values may not have any substantial amount of such oxides. This evident from an early research on P fixation by Kanwar and Grewal (1960) who reported about 30% P fixation capacity of calcareous and alkali soils due to the presence of free sesquioxides. On the other hand Velayutham et al. (2002) reported the presence of substantial amount of P in alkali soils and thus such soils responded to added P after few years of initial cropping with rice and wheat. Although work on characterization and estimation of amorphous constituents in soils of India was intensified in 1970s, proper methods for their characterization and for obtaining data for verifying the effects on soil properties are still awaited (Krishna Murti 1982).

8.2.4 Potassium

8.2.4.1 Vertisols

Vertisols are stated to be adequately supplied with potassium (K), and therefore, responses to applied K are generally not obtained (Pal et al. 2012b). Extensive research on K behaviour in Indian Vertisols for the last two-and-a-half decades may be a good example for understanding the basic issues of K adsorption and desorption (Pal 2003). As the Deccan basalt does not contain micas (Pal and Deshpande 1987), the Vertisols derived from its alluvium contain small amounts of micas, which are concentrated mainly in their silt and coarse clay fractions (Pal et al. 2001). Micas are added to Vertisols during the erosional and depositional episodes experienced by the Deccan basalt areas during the post-Plio-Pleistocene transition period (Pal and Deshpande 1987). Petrographic and scanning electron microscope (SEM) examinations of the muscovites and biotites of the Vertisols of Peninsular India indicate little or no alteration (please refer to Fig. 2.2a, b) (Pal et al. 2001; Srivastava et al. 2002). Therefore, highly available K status of Vertisols is related to the retention of elementary layers of the micas, which favours the release of K^+ even though its content is low. The precise nature of silt and clay sized micas was determined on the basis of the X-ray intensity ratio of peak heights of 001 and 002 basal reflections of 1.0 nm minerals (micas) (Pal et al. 2001). The ratio is generally greater than unity in the silt but is close to unity in the clay fraction (please refer to Fig. 2.2c). The ratio >1 suggests the presence of muscovite and biotite minerals. If muscovite minerals were present alone, the ratio would have been close to unity. In the event of a mixture of these two micas, both will contribute to the intensity of the 1.0 nm reflections, whereas the contribution of biotite to the 0.5 nm reflection would be nil or negligible, thus giving a higher value to the intensity ratio of these reflections (Pal et al. 2001). Thus, the silt fractions of the soils contain both muscovite and biotite; whereas the clay fractions are more muscovitic in character (please refer to Fig. 2.2c). The enrichment of soils with muscovite is not favourable so far as the K release is concerned. This is evidenced by the reduced rate of K release in the Vertisols compared with the much higher rate of K release from the biotite-enriched soils of the Indo-Gangetic Plains (IGP) (Pal et al. 2001) when they were subjected to repeated batch type Ba–K exchange under identical experimental conditions (Pal and Durge 1989). Muscovite and biotite micas co-exist in soil environments. The weathering of muscovite in the presence of biotite is improbable. Therefore, the quantity of muscovite cannot be used as an index of K reserve in soils (Pal et al. 2001). For this reason, a selective quantification of biotite mica in the common situation in which soils contain mixtures of biotite and muscovite was envisaged (Pal et al. 2001). The contents of biotite in Vertisols and their size fractions were estimated through a rigorous and exhaustive Ba-K exchange reaction. The cumulative amount of K released at the end of final extraction by the soil's size fractions when the release of K nearly ceased was considered as mainly coming from biotite (Pal et al. 2006b) (Fig. 8.1). The amount of clay biotite, silt biotite and sand biotite in the representative Vertisols of central India ranged from 1.0 to 1.6, 0.2 to 0.3 and 0.2 to 0.4%, respectively, constituting 7–19, 2–3 and 2–5% of the total mica in the respective size fractions. In the <2 mm fine earth fraction, the biotite quantity does not exceed 1%, which constitutes approximately 6–8% of the total mica. For any size fraction, the cumulative amount of K released on a biotite weight basis follows the order: cumulative amount of K released on the entire mica weight basis > cumulative amount of K released on the

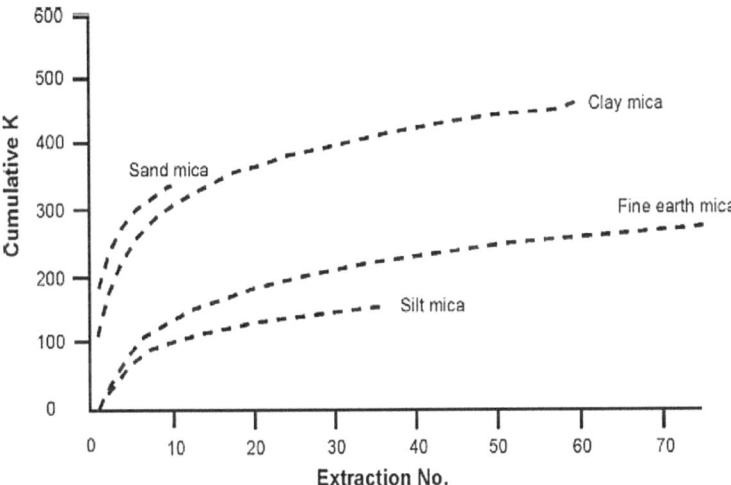

Fig. 8.1 Relationship between numbers of extractions and cumulative K release (mg/100 g^{-1} mica) of micas in various size fractions of a Vertisol (Adapted from Pal et al. 2006b)

weight basis of the size fraction (Table 8.2). The significant positive correlation between the cumulative K release of soils and their size fractions is mainly from biotite and is established from the statistical analysis of bivariate data sets of several parameters that directly or indirectly influence K release. The significant positive correlations between cumulative K release from sand, silt and clay and their corresponding total K contents, respectively (Table 8.3), indicate that the K release is a function of total K content in micas and feldspars. However, the positive correlations between total K contents in sand, silt, clay and soil and their mica contents

Table 8.2 Cumulative K release from a representative Vertisol and its size fractions

Horizon	Depth (cm)	Fine earth (<2 mm) cumulative K release in 75 extractions			Sand (2–0.05 mm) cumulative K release in 10 extractions			Silt (0.05–0.002 mm) cumulative K release in 35 extractions			Clay (<0.002 mm) cumulative K release in 60 extractions		
		SF[a]	MB	BB	SF	MB	BB	SF	MB	BB	SF	MB	BB
		mg K 100 g^{-1}											
Ap	0–15	69	429	6059	20	272	7000	16	191	7004	114	561	6990
Bw1	15–41	41	277	4230	12	162	7006	15	195	7009	92	509	6998
Bw2	41–70	39	267	4097	23	297	6997	13	184	7011	88	502	6999
Bss1	70–95	45	261	4638	15	191	6986	14	161	6990	91	433	7000
Bss2	95–135	49	286	4793	24	334	6991	15	162	6991	92	462	6999
Bss3	135–155	37	235	3849	13	147	6907	16	184	7008	94	471	6984

Adapted from Pal et al. (2006b)

[a]*SF* On the basis of size fraction; *MB* On the basis of mica content; *BB* On the basis of biotite content

(Table 8.3) indicate the predominant influence of mica to supply K to the plants grown in Vertisols. Furthermore, significant positive correlations between the cumulative K release of sand, silt, clay and soil and their respective mica contents (Table 8.3) indicate that the K release from either the soils or different size fractions are controlled mainly by mica. However, better correlations than those between the cumulative K release of sand, silt, clay and soil and their biotite contents (Table 8.3) provide incontrovertible evidence that the K release in soils is primarily controlled by biotite mica. This further supports the earlier observations on the inertness of muscovite mica in the release of K in the presence of biotite (Pal et al. 2001). The released amount of K from sand-, silt- and clay-sized biotite (Table 8.3) is in contrast to the relationships observed between cumulative K release and particles of specimen biotite by earlier researchers (Pal 1985; Pal et al. 2001). This indicates that large-sized biotite particles have a lower K selectivity than finer particles. Comparable cumulative amounts of K released from sand and silt biotite (Table 8.3) indicate that not only the sand-sized but also some portion of silt-sized biotite of the Vertisols have a greater number of elementary layers along with little weathered biotite. It is observed that during the formation of Vertisols since the Holocene (Pal et al. 2009), there has been no substantial weathering of biotite under the SAT environments. This validates the earlier hypothesis (Srivastava et al. 2002) that the formation of Vertisols reflects a positive entropy change due to a lack of any substantial weathering of primary minerals. The relevance of the almost-unaltered biotites (please refer to Fig. 2.2b) is that both sand and silt biotites have highly favourable K release potential, which is reflected in the medium to highly available K status of the Vertisols of India. Agronomic experiments on the Vertisols of central India have indicated crop response to K fertilizers after two

Table 8.3 Co-efficient of correlation among various soil characteristics

Parameter		r
Cumulative K of sand	Total K in sand	0.635**
Cumulative K of silt	Total K in silt	0.771**
Cumulative K of clay	Total K in clay	0.822**
Total K in sand	Sand mica	0.933**
Total K in silt	Silt mica	0.766**
Total K in clay	Clay mica	0.981**
Total K in soil	Soil mica	0.979**
Cumulative K of sand	Sand mica	0.524*
Cumulative K of silt	Silt mica	0.694**
Cumulative K of clay	Clay mica	0.851**
Cumulative K of soil	Soil mica	0.429*
Cumulative K of sand mica	Sand biotite	0.894**
Cumulative K of silt mica	Silt biotite	0.917**
Cumulative K of clay mica	Clay biotite	0.978**
Cumulative K of soil mica	Soil biotite	0.435*

Adapted from Pal et al. (2006b)
*Significant at 0.05 level; **Significant at 0.01 level

years of cropping with hybrid cotton (Pal and Durge 1987). Therefore, the present available K status may not be sustainable over a longer term because the contents of sand and silt biotites are low. This information helps dispel the myth that the Vertisols are rich in available K and that they may not warrant the application of K fertilizers.

Potassium adsorption/fixation in Vertisols does not appear to be sufficiently severe to conclude that K becomes completely unavailable to plants. The study by Pal and Durge (1987) on K adsorption by the Vertisols of Peninsular India indicates that fine clay smectites adsorbed 50–60% of added K, amounting 25–30 mg K/100 g clay. This apparently suggests that the fine clay smectites of Indian Vertisols are close to beidellite (Bajwa 1980). Through a series of diagnostic methods to characterize the fine clay smectites, Pal and Deshpande (1987) confirmed that they are nearer to the montmorillonite of the montmorillonite–nontronite series. Because smectites can have no K selectivity (Brindley 1966), further characterization of fine clay smectites (Pal and Durge 1987) indicated the presence of vermiculites, which are generally not detected on the glycolation of Ca-saturated samples but can be detected by a progressive reinforcement of the 1.0 nm peak of mica while heating the K saturated samples from 25 to 550 °C (please refer to Fig. 2.7). The content of vermiculite was quantified following the method of Alexiades and Jackson (1965) by Pal and Durge (1987), and the vermiculite content ranged from 5 to 9% in the fine clay of Vertisols. Pal and Durge (1987) concluded that the observed K adsorption by the silt and clay fractions is due to the presence of vermiculite and not to smectite. The smectites of Vertisol clays belong to the low-charge dioctahedral type, and thus, they expand beyond 1.0–1.4 nm with the glycolation of K-saturation and heating the samples at 300 °C (please refer to Fig. 2.7). These smectites, when treated according to the alkylammonium method (Lagaly 1994), showed the presence of both monolayer to bilayer and bilayer to pseudotrilayer transitions. The layer charge of the half-unit cell of smectite ranges from 0.28 to 0.78 $mol(-)/(SiAl)_4O_{10}(OH)_2$, and the low-charge smectite constitutes >70% in them (Ray et al. 2003). The position of the higher charge with 0.78 units or lower appears to be due to the presence of small amounts of vermiculite as determined quantitatively by Pal and Durge (1987). The limited leaching in Vertisols and small amount of vermiculite would lessen the rate of added K-fertilizers when required. If K fertilizers are added as a basal dose, the K^+ ions would be fixed in the interlayer of vermiculite, which would make the NH_4 ions from N fertilizers more labile for ready assimilation by growing plants if not added as a basal dose.

Vertisols developed on the zeolitic Deccan basalt or in its alluvium, are also adequately supplied with K as evident from their high to very high available K status even in the subsurface (Table 8.4). Potassium release in soils is primarily controlled by biotite mica, which constitutes only about 1% in the <2 mm fine earth fraction in Vertisols. The apparent incompatibility between medium to high

Table 8.4 Selected soil properties of zeolitic Chromic/Sodic Haplusterts cultivated to rice crops

Depth (cm)	pH (1:2)	ECe (dS m^{-1})	CaCO$_3$ (<2 mm) (%)	Organic carbon (%)	ESP	sHC[a] (mm h^{-1})	Base saturation (%)	Available K (kg ha^{-1})
Sakka soils—Chromic Haplusterts								
0–15	5.2	0.26	1.6	0.73	0.9	18	93	429
15–34	5.3	0.16	1.7	0.54	0.9	36	109	343
34–59	5.3	0.26	1.9	0.40	0.9	35	106	343
59–93	5.4	0.10	2.0	0.38	1.1	10	93	343
93–141	7.3	0.12	4.1	0.39	1.2	15	115	429
141–155	7.9	0.21	10.0	0.22	0.8	16	107	343
Teligi soils—Sodic Haplusterts								
0–10	7.9	0.4	10.5	1.55	1.5	62	103	515
10–25	8.0	0.3	10.7	0.81	1.7	27	113	343
25–44	8.0	0.4	12.2	0.76	1.8	29	109	343
44–69	7.8	0.4	10.3	0.73	4.0	21	117	257
69–97	7.6	0.3	5.9	0.69	3.4	11	108	515
97–123	8.6	0.4	15.1	0.50	16.8	3	110	257

Adapted from Pal et al. (2003); [a]23 mm h^{-1} is the weighted mean sHC in 0–100 cm depth of Sakka soils and 24 mm h^{-1} is the weighted mean sHC in 0–100 cm depth of Teligi soils

ECe Electrical conductivity of the saturation extract; *ESP* exchangeable sodium percentage; *sHC* saturated hydraulic conductivity

available K status in surface horizons (250 kg K ha^{-1}; Table 8.4) and low biotite mica in soils of the Deccan basalt areas, need further insight in view of pronounced selectivity of zeolites for K$^+$ and NH$_4^+$ ions. Contribution of zeolites to available K of soils is not uncommon (Brown et al. 1969; Talibudeen and Weir 1972).

Potassium adsorption/fixation in Vertisols has been attributed to the presence of vermiculite only (Pal et al. 2012b). However, in the presence of zeolites K adsorption should not be totally attributed to vermiculite as zeolites also have strong selectivity for K$^+$ ion. In view of the role of zeolites in adsorption and desorption of K$^+$ and NH$_4^+$ ions alongside vermiculite, a fresh research initiative is warranted to pinpoint the selective contribution of zeolite, biotite and vermiculite on a time scale when they co-exist in soil environments.

8.2.4.2 RF Soils

Potassium (K) is also one of the limiting nutrient elements in RF soils of India, especially in the HT climate areas (Pal et al. 2001). In micaceous soils of HT climate in NEH areas, crops respond to added K up to 80 kg K ha^{-1} (Datta 2013) because the soils are rich in sand size muscovite mica (Fig. 8.2a). K release from muscovite is inhibited in the presence of biotite (Pal et al. 2001). The X-ray intensity ratio of peak heights of 001 and 002 basal reflections of mica in the silt

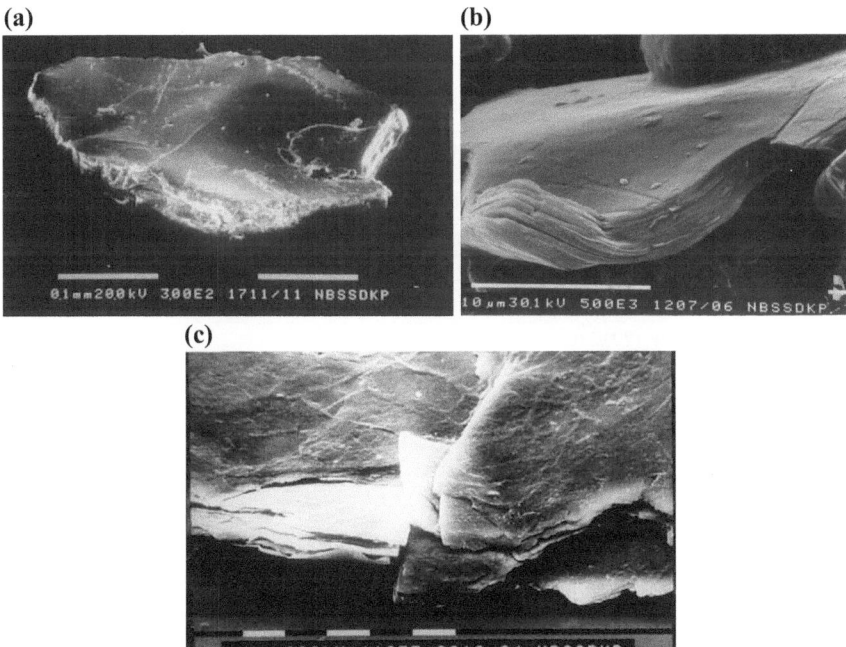

Fig. 8.2 Representative SEM photographs of unaltered muscovites in acidic Inceptisols (**a**), Dahotia soils, Assam of HT climate, and almost unaltered biotites with minor layer separation and bending of edges in SAT Alfisols (**b**), Patancheru, Andhra Pradesh and (**c**), Dyavapatna, Karnataka (Adapted from Pal et al. 2000a, 2001)

fraction of acidic Alfisols (Chandran et al. 2004), Ultisols (Chandran et al. 2005) and acidic Inceptisols (Pal et al. 1987) is greater than unity; and that of the clay fraction is close to unity (Table 8.5), indicating that silt fractions contain both muscovite and biotite, and the clay fraction is more muscovitic in character (Pal et al. 2001). The inertness of muscovite in releasing K makes K a limiting nutrient in RF soils of HT climate.

In a repeated batch type experiment with Ba–K exchange, K release from these soils (Dahotia soils of Assam in NEH, Pal and Durge 1989) was reduced to almost nil after a few extractions, indicating rather slow release of K from muscovite. Such highly weathered soils fix relatively high amount of added K (Fox 1982; Pal and Durge 1989) because of the presence of vermiculite in their silt (20–6 μm, and 6–2 μm) and coarse clay (2–0.6 μm) fractions, and low charge vermiculite in their medium (0.6–0.2 μm) and fine clay (<0.2 μm) fractions (Pal and Durge 1989). The fine clay of Dahotia soils fixes the most amount of the added K (Fig. 8.3.).

The predominance of kaolinite followed by illite is generally believed to be reason for the low nutrient holding capacity of RF soils (Alfisols) of the southern Peninsular India under SAT environment; and therefore application of balanced fertilizer including K is generally recommended. It is interesting to note that in

Table 8.5 X-ray intensity ratio of the peak heights of 001/002 basal reflection in the silt and clay fractions

Benchmark soil/soil series	Parent material	Size fractions	
		50–2 μm	<2 μm
Dahotia (Assam-HT climate) (Typic Haplaquept)	Alluvium	1.47	1.10
Akahugaon (Assam-HT climate) (Typic Haplaquept)	Alluvium	1.70	1.04
Patancheru (RF soil-SAT) (Udic Rhodustalf)	Granite-Gneiss	1.77	1.80
Nalgonda (RF soil-SAT) (Udic Rhodustalf)	Granite-Gneiss	2.00	1.87
Dyavapatna (RF soil-SAT) (Udic Rhodustalf)	Granite-Gneiss	2.25	2.16

Adapted from Pal et al. (2012a)

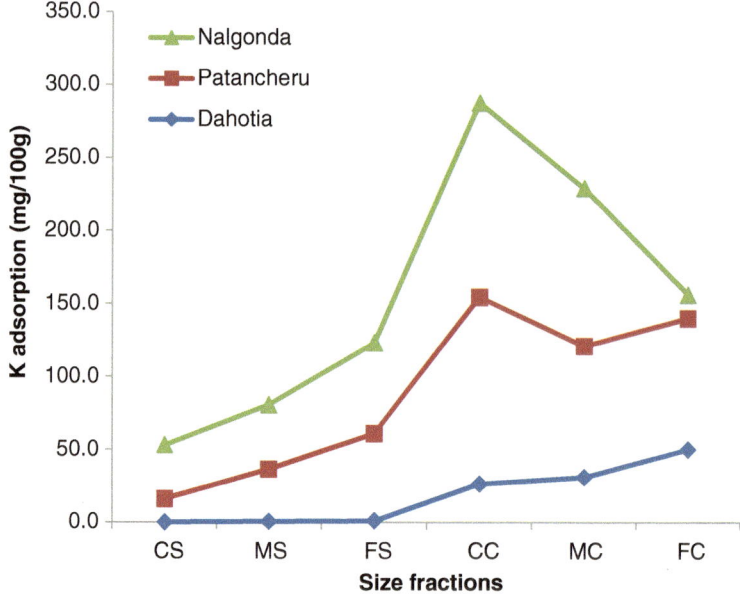

Fig. 8.3 Potassium adsorption by various size fractions of RF Alfisols of the SAT regions (Patancheru and Nalgonda, Andhra Pradesh, India) and acidic Inceptisols (Dahotia of the HT region in Assam, India). *CS* coarse silt (50–20 μm), *MS* medium silt (20–6 μm), *FS* fine silt (6–2 μm), *CC* coarse clay (2–0.6 μm), *MC* medium clay (0.6–0.2 μm), *FC* fine clay (<0.2 μm) (Adapted from Pal et al. 1993 and Pal and Durge 1989)

many of these soils crops do not respond to K fertilizer application (Pal et al. 2014). The X-ray intensity ratio of peak heights of 001 and 002 basal reflections of mica in the silt and clay fractions of many SAT Alfisols is near 2 (Patancheru, Nalgonda and Dyavapatna soils, Table 8.5), suggesting that these soils contain sufficient biotite.

The RF Alfisols of the SAT are relict paleosols but polygenetic in nature. Due to climate change from tropical humid to semi-arid during the Plio-Pleistocene transition period, the upper layers of the soils of the preceding tropical humid climate were truncated by multiple arid erosional cycles, which exposed the relatively less weathered lower layers wherein considerable amount of unaltered biotite particles remained in the sand and silt fractions (Fig. 8.2b, c) (Pal et al. 1989, 1993). Many of the proposed relationships between K release and mica particle size hitherto obtained either with soil or specimen minerals are, therefore, not valid in these polygenetic soils (Pal et al. 1993).

The cumulative K release from different size fractions of two benchmark RF soils (Alfisols) of SAT namely Patancheru and Nalgonda of southern India, showed a fairly high rate of K release (Pal et al. 1993) but the observed data indicated a contrasting particle size-K release relationship between silts and clays. K release increased with the fineness of clay size mica while it decreased with the fineness of silt size mica (Fig. 8.4). The zones in clay micas are at different stages of expansion unlike those of the silt mica, and under such circumstances K release would occur mainly by edge weathering as is evident from increased K release with the decrease in clay mica particle (Pal and Durge 1989).

K release increased with the increase in particle size of silt sized mica, which are almost unweathered sand and silt biotites (Fig. 8.2b, c). This particular K release trend is normally obtained with specimen micas (Pal 1985). Thus, quite favourable K release rate from both silt and clay micas explains as to why crop response to fertilizer K is seldom obtained in many of RF Alfisols under SAT environments.

It is often reported that micas, hydrous micas and vermiculites have high adsorption/fixation properties. But mica indeed does not expand on being saturated with divalent cations and unlikely adsorbs added K (Sarma 1976). In such reactions interlayer charge density of the mineral is of fundamental importance. Kaolinites are of no significance in such a reaction, while vermiculites are converted to mica by layer contraction by K. Smectites would not possess this property as their layer charge is too low and they do not adsorb K selectively (Rich 1968b) unless the charge density is high like in high charge smectite or low charge vermiculite (Pal and Durge 1989).

Due to impoverishment of vermiculite, the observed K adsorption by silt and clay fractions of benchmark RF Alfisols of Patancheru and Nalgonda series (Fig. 8.3) may be attributed to smectite content because it increases with the decrease in soil size fractions (Pal et al. 1993). However, perusal of the K adsorption data (Fig. 8.3) indicates an incompatibility between the amount of clay smectite and the extent of K adsorption by clay fractions. Despite sufficient amount of smectite, the finer fractions of clay particularly the fine clay fractions (<0.2 μm) does not adsorb K proportionately. Soil clays contain both low and high charge smectites and their co-existence is, however, related to their respective genesis in paleoclimatic environments (Pal et al. 1989). Pal et al. (1989) demonstrated that considerable amount of well crystallized clay size low charge dioctahedral smectite

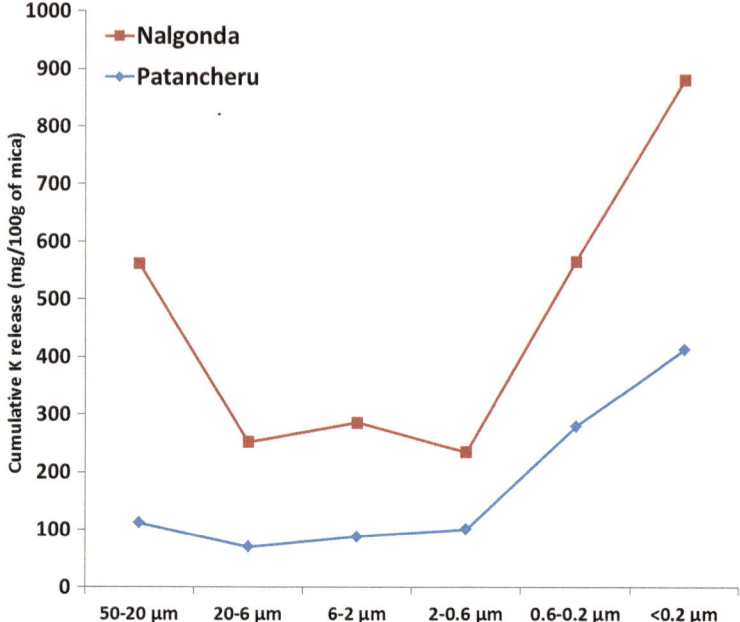

Fig. 8.4 Potassium release from various size fractions of RF Alfisols (Patancheru and Nalgonda) of southern India *CS* coarse silt (50–20 μm), *MS* medium silt (20–6 μm), *FS* fine silt (6–2 μm), *CC* coarse clay (2–0.6 μm), *MC* medium clay (0.6–0.2 μm), *FC* fine clay (<0.2 μm) (Adapted from Pal et al. 1993)

was the first weathering product of granite-gneiss that survived transformation to kaolinite in a Pre-Pliocene tropical humid climate and is preserved to the present along with kaolinite, which does not participate in K adsorption.

The biotite which somehow survived HT climate weathering, altered to trioctahedral smectite in the silt and coarse clay fractions under the semi-arid climate of the Plio-Pleistocene transition period. The results on the selective K adsorption by trioctahedral smectite and not by dioctahedral smectite are in accord with those reported earlier with high and low charge smectites respectively (Pal and Durge 1987, 1989). SAT Alfisols are generally rich in clay, especially fine clays and therefore, it is envisaged that these soils may have less K adsorption/fixation problem (Pal et al. 2000a).

8.2.4.3 IGP Soils

Alluvial soils, one of the major soil groups in India, are important with respect to agricultural potential. These soils contain a good amount of sand and silt size micas, especially biotites (please refer to Fig. 4.7). These micaceous soils are rich in available K, and the K applied to many benchmark soils undergoes fixation, indicating the presence of 2:1 minerals capable of fixation of added K. Crops in benchmark soils of semi-arid climate seldom respond to K-fertilizer even under continuous cropping over long periods of time (Pal and Mondal 1980). But, in soils of per humid climate crops respond to K-fertilizers (Roy et al. 1978) even though soils contain a considerable amount of muscovite with subordinate amount of biotite (Pal et al. 1987).

Much of the investigative work on potassium and mineralogy is related to specimen minerals and therefore, many of the proposed relationships are speculative (Rich 1972; Sarma 1984). In order to gather precise information on the K release and adsorption reactions of the IGP benchmark soils in relation to the specific nature of soil minerals Pal and Durge (1989) made an extensive study.

Following the repeated batch type for Ba-K exchange, Pal and Durge (1989) reported the K release from soil mica increased with a decrease in soils' particle size, and the rate of K release was dependent upon the nature of soil mica, biotite mica in particular (Pal et al. 2001). The much reduced rate of K release with higher number of extractions was due to more of muscovite character of soil mica, suggesting that muscovite was a rather useless source of K reserve. The observed negative correlation between K release and particle size, a trend opposite to that of specimen minerals (Reichenbach 1972; Pal 1985), is explained on the basis of edge weathering of soil mica. Zones in soil micas are expected to contain layer minerals in different stages of expansion and this fact is evidenced by the observed interstratification of mica and expanded layers found upon X-ray examination of silt and clay fractions of soils (Pal and Durge 1989). Under such circumstances K release occurs mainly by edge weathering of soil mica as evident from the higher K release with the decrease in particle size (Fannings and Keramidas 1977). The slower release by coarser, as opposed to finer mica particles, by edge weathering could be best explained as suggested by Reed and Scot (1962). These authors explained this phenomenon in terms of (i) the smaller peripheral surface with larger particles, per unit weight of material, across which K ions may diffuse, and (ii) the greater distance that K ions must diffuse, with larger particles, for a given fraction of K removed.

Vermiculite as defined by Alexiades and Jackson (1965) is responsible for the adsorption of K. The observed selective adsorption of K by smectite (Pal and Durge 1989), is a fact in contrast to the behaviour of specimen mineral (Rich 1968a, b) and soil clay smectite of dioctahedral nature (Pal and Durge 1987). The higher correlation coefficient at 1% level of significance (r = 0.899) between the K adsorption value (mg/100 g) and the vermiculite plus smectite content (g/100 g) of the silt and the clay fractions, as compared to only the vermiculite content (r = 0.725) does suggest that clay smectites also adsorbed K selectively. The smectite, as defined according to Alexiades and Jackson (1965), expanded to 17 Å peak on glycolation but contracted readily to 1.0 nm on K-saturation at 110 °C indicating its high layer charge density and thus demonstrates the critical role of layer charge density in the adsorption of K by clay smectites of the many IGP soils developed in the micaceous alluvium of Himalayan origin. Research results obtained on the nature of mica and smectite and their fundamental relation to K release and fixation have important implications in K management of the vertic and non-vertic IGP soils (Pal et al. 2010).

References

Abrol IP, Gupta RK (1998) Indo-Gangetic plains: issues of changing land use. LUCC Newsl 1:5 (March 1998, no. 3)

Ae N, Arihara J, Okade K, Yoshihara TC, Johansen C (1990) Phosphorus uptake by pigeonpea and its role in cropping systems of the Indian subcontinent. Science 248:477–480

Alexiades CA, Jackson ML (1965) Quantitative determination of vermiculite in soils. Soil Sci Soc Am Proc 29:522–527

Bajwa I (1980) Soil clay mineralogies in relation to fertility management: effect of soil clay mineral compositions on potassium fixation under conditions of wetland rice culture. Commun Soil Sci Plant Anal 11:1019–1027

Bhatia A, Pathak H, Aggarwal P (2004) Inventory of methane and nitrous oxide emissions from agricultural soils of India and their global warming potential. Curr Sci 87:317–324

Bhatia A, Aggarwal PK, Jain N, Pathak H (2012) Greenhouse gas emission from rice–wheat growing areas in India: spatial analysis and upscaling. Greenhouse Gases Sci Technol 2: 115–125

Bhattacharyya T, Sen TK, Singh RS, Nayak DC, Sehgal JL (1994) Morphology and classification of Ultisols with Kandic horizon in north eastern region. J Indian Soc Soil Sci 42:301–306

Bhattacharyya T, Pal DK, Velayutham M, Chandran P, Mandal C (2000a) Total carbon stock in Indian soils: issues, priorities and management. Land resource management for food and environmental security. Soil Conservation Society of India, New Delhi, pp 1–46

Bhattacharyya T, Pal DK, Srivastava P (2000b) Formation of gibbsite in presence of 2:1 minerals: an example from Ultisols of northeast India. Clay Miner 35:827–840

Bhattacharyya T, Pal DK, Chandran P, Mandal C, Ray SK, Gupta RK, Gajbhiye KS (2004) Managing soil carbon stocks in the Indo-Gangetic Plains, India. Rice–wheat consortium for the Indo-Gangetic Plains, New Delhi, p 44, (http://www.rwc-prism.cgiar.org and http://www.cimmyt.org)

Bhattacharyya T, Pal DK, Chandran P, Ray SK (2005) Land-use, clay mineral type and organic carbon content in two Mollisols–Alfisols–Vertisols catenary sequences of tropical India. Clay Res 24:105–122

Bhattacharyya T, Pal DK, Lal S, Chandran P, Ray SK (2006) Formation and persistence of Mollisols on zeolitic Deccan basalt of humid tropical India. Geoderma 136:609–620

Bhattacharyya T, Pal DK, Easter M, Batjes NH, Milne E, Gajbhiye KS, Chandran P, Ray SK, Mandal C, Paustian K, Williams S, Killian K, Coleman K, Falloon P, Powlson DS (2007a) Modelled soil organic carbon stocks and changes in the Indo-Gangetic Plains, India from 1980 to 2030. Agric Ecosyst Environ 122:84–94

Bhattacharyya T, Chandran P, Ray SK, Pal DK, Venugopalan MV, Mandal C, Wani SP (2007b) Changes in levels of carbon in soils over years of two important food production zones of India. Curr Sci 93:1854–1863

Bhattacharyya T, Sarkar D, Sehgal JL, Velayutham M, Gajbhiye KS, Nagar AP, Nimkhedkar SS (2009) Soil taxonomic database of India and the States (1:250,000 scale), NBSSLUP Publ. 143, NBSS&LUP, Nagpur, India, (266 pp)

Bhattacharyya T, Chandran P, Ray SK, Mandal C, Tiwary P, Pal DK, Wani SP, Sahrawat KL (2014) Processes determining the sequestration and maintenance of carbon in soils: a synthesis of research from tropical India. Soil Horizons, Published 9 July 2014, pp 1–16. doi:10.2136/sh14-01-0001

Bhattacharyya T, Chandran P, Pal DK, Mandal C, Mandal DK (2015) Distribution of zeolitic soils in India. Curr Sci 109:1305–1313

Brindley GW (1966) Ethylene glycol and glycerol complexes of smectites and vermiculites. Clay Miner 6:237–259

Brown G, Catt JA, Weir AH (1969) Zeolites of the clinoptilolite–heulandite type in sediments of south-east England. Miner Mag 37:480–488

Burford JR, Sahrawat KL (1989) Nitrogen availability in SAT soils: environment effects on soil processes. Soil fertility and fertility management in semi-arid tropical India. In: Christianson CB (ed) Proceedings Colloquim held at ICRISAT Centre, Patancheru, India, Oct 10–11, 1988. International Fertilizer Development Centre, Muscle Shoals, AL, pp 53–60

Chandran P, Ray SK, Bhattacharyya T, Dubey PN, Pal DK, Krishnan P (2004) Chemical and mineralogical characteristics of ferruginous soils of Goa. Clay Res 23:51–64

Chandran P, Ray SK, Bhattacharyya T, Srivastava P, Krishnan P, Pal DK (2005) Lateritic soils of Kerala, India: their mineralogy, genesis and taxonomy. Aust J Soil Res 43:839–852

Dalal RC (1977) Fixed ammonium and carbon–nitrogen ratios of Trinidad soils. Soil Sci 124:323–327

Datta M (2013) Soils of north-eastern region and their management for rain-fed crops. In: Bhattacharyya T, Pal DK, Sarkar D, Wani SP (eds) Climate change and agriculture. Studium Press, New Delhi, pp 19–50

Dhillon SK, Dhillon KS (1991) Characterization of potassium in red (alfisols), black (vertisols) and alluvial (inceptisols and entisols) soils of India using electro-ultra filtration. Geoderma 50:185–196

Fannings DS, Keramidas VZ (1977) Micas. In: Dixon JB, Weed SB (eds) Minerals in soil environments. Soil Science Society of America, Wisconsin, USA, pp 195–258

Fox RL (1982) Some highly weathered soils of Puerto Rico, 2. Chemical properties. Geoderma 27:139–176

Ghosh SK (1997) Clay research in India. 13th Prof. J.N. Mukherjee ISSS Foundation Lecture. J Indian Soc Soil Sci 45:637–658

Haynes RJ (1983) Effect of lime and phosphate applications on the adsorption of phosphate, sulphate, and molybdate by a spodosol. Soil Sci 135:221–226

ICRISAT (1988) Phosphorus in Indian vertisols. Summary Proceedings of a Workshop, August 1988, ICRISAT Center, Patancheru, A.P. 502324, India, pp 23–26

Jangra R, Gupta SR, Singh N (2015) Plant biomass, productivity, and carbon storage in an ecologically restored grassland on a sodic soils in north-western India. Indian J Sci 20:85–96

Kanwar JS, Grewal JS (1960) Phosphate fixation in Punjab soils. J Indian Soc Soil Sci 8:211–218

Krishna Murti GSR (1982) Amorphous constituents of soil clays. In: Review of soil research in India. Transactions of the 12th International Congress of Soil Science vol 2, pp 723–730

Lagaly G (1994) Layer charge determination by alkylammonium ions. Layer charge characteristics of 2:1 silicate clay minerals. In: Mermut AR (ed) CMS workshop lectures, vol 6. The Clay Minerals Society, Boulder, USA, pp 1–46

Majumder B, Mandal B, Bandyopadhyay PK, Chaudhury J (2007) Soil organic pools and productivity relationships for a 34 year old rice–wheat–jute agroecosystem under different fertilizer treatments. Plant Soil 297:53–67

Mandal B, Majumder B, Bandyopadhyay B, Hazra GC, Gangopadhyay A, Samantaroy RN, Misra AK, Chowdhuri J, Saha MN, Kundu S (2007) The potential of cropping systems and soil amendments for carbon sequestration in soils under long-term experiments in subtropical India. Glob Change Biol 13:357–369

Mandal B, Majumder B, Adhya TK, Bandyopadhyay PK, Gangopadhyay A, Sarkar D, Kundu MC, Gupta Choudhury S, Hazra GC, Kundu S, Samantaray RN, Misra AK (2008) Potential of double-cropped rice ecology to conserve organic carbon under subtropical climate. Glob Change Biol 14:1–13

Ming DW, Allen ER (2001) Use of natural zeolites in agronomy, horticulture and environmental soil remediation. Rev Miner Geochem 45:619–654

Newman ACD (1984) The significance of clays in agriculture and soils. Philos Trans Roy Soc Lond A311:375–389

Nommik H, Vahtras K (1982) Retention and fixation of ammonium and ammonia in soils. In: Stevensen FJ (ed) Nitrogen in agricultural soils. Agronomy 22:123–171

Pal DK (1985) Potassium release from muscovite and biotite under alkaline conditions. Pedologie 35:133–146

Pal DK (2003) Significance of clays, clay and other minerals in the formation and management of Indian soils. J Indian Soc Soil Sci 51:338–364

Pal DK, Mondal RC (1980) Crop response to potassium in sodic soils in relation to potassium release behaviour in salt solutions. J Indian Soc Soil Sci 26:347–354

Pal DK, Deshpande SB (1987) Genesis of clay minerals in a red and black complex soils of southern India. Clay Res 6:6–13

Pal DK, Durge SL (1987) Potassium release and fixation reactions in some benchmark Vertisols of India in relation to their mineralogy. Pedologie 37:103–116

Pal DK, Durge SL (1989) Release and adsorption of potassium in some benchmark alluvial soils of India in relation to their mineralogy. Pedologie (Ghent) 39(235):248

Pal DK, Deshpande SB, Durge SL (1987) Weathering of biotite in some alluvial soils of different agro climatic zones. Clay Res 6:69–75

Pal DK, Deshpande SB, Venugopal KR, Kalbande AR (1989) Formation of di and trioctahedral smectite as an evidence for paleoclimatic changes in southern and central Peninsular India. Geoderma 45:175–184

Pal DK, Deshpande SB, Durge SL (1993) Potassium release and adsorption reactions in two ferruginous (polygenetic) soils of southern India in relation to their mineralogy. Pedologie (Ghent) 43:403–415

Pal DK, Bhattacharyya T, Deshpande SB, Sarma VAK, Velayutham M (2000a) Significance of minerals in soil environment of India, NBSS Review Series 1. NBSS&LUP, Nagpur 68p

Pal DK, Dasog GS, Vadivelu S, Ahuja RL, Bhattacharyya T (2000b) Secondary calcium carbonate in soils of arid and semi-arid regions of India. In: Lal R, Kimble JM, Eswaran H, Stewart BA (eds) Global climate change and pedogenic carbonates. Lewis Publishers, Boca Raton, Fl, pp 149–185

Pal DK, Srivastava P, Durge SL, Bhattacharyya T (2001) Role of weathering of fine grained micas in potassium management of Indian soils. Appl Clay Sci 20:39–52

Pal DK, Bhattacharyya T, Ray SK, Bhuse SR (2003) Developing a model on the formation and resilience of naturally degraded black soils of the peninsular India as a decision support system for better land use planning. NRDMS, Department of Science and Technology (Govt. of India) Project Report, NBSSLUP (ICAR), Nagpur, 144 pp

Pal DK, Bhattacharyya T, Ray SK, Chandran P, Srivastava P, Durge SL, Bhuse SR (2006a) Significance of soil modifiers (Ca-zeolites and gypsum) in naturally degraded Vertisols of the Peninsular India in redefining the sodic soils. Geoderma 136:210–228

Pal DK, Nimkar AM, Ray SK, Bhattacharyya T, Chandran P (2006b). Characterisation and quantification of micas and smectites in potassium management of shrink–swell soils in Deccan basalt area. In: Benbi DK, Brar MS, Bansal SK (eds) Balanced fertilization for sustaining crop productivity. Proceedings of the International Symposium held at PAU, Ludhiana, India, 22–25 Nov 2006 IPI, Switzerland, pp 81–93

Pal DK, Bhattacharyya T, Chandran P, Ray SK, Satyavathi PLA, Durge SL, Raja P, Maurya UK (2009) Vertisols (cracking clay soils) in a climosequence of Peninsular India: evidence for Holocene climate changes. Quatern Int 209:6–21

Pal DK, Lal S, Bhattacharyya T, Chandran P, Ray SK, Satyavathi PLA, Raja P, Maurya UK, Durge SL, Kamble GK (2010) Pedogenic thresholds in Benchmark soils under rice-wheat cropping system in a climosequence of the Indo-Gangetic Alluvial Plains. Final Project Report, Division of Soil Resource Studies. ICAR-NBSS & LUP, Nagpur (193 pp)

Pal DK, Bhattacharyya T, Chandran P, Ray SK (2012a) Linking minerals to selected soil bulk properties and climate change: a review. Clay Res 31:38–69

Pal DK, Wani SP, Sahrawat KL (2012b) Vertisols of tropical Indian environments: pedology and edaphology. Geoderma 189–190:28–49

Pal DK, Bhattacharyya T, Sinha R, Srivastava P, Dasgupta AS, Chandran P, Ray SK, Nimje A (2012c) Clay minerals record from late quaternary drill cores of the Ganga Plains and their implications for provenance and climate change in the Himalayan Foreland. Palaeogeogr Palaeoclimatol Palaeoecol 356–357:27–37

Pal DK, Wani SP, Sahrawat KL (2013) Zeolitic soils of the Deccan basalt areas in India: their pedology and edaphology. Curr Sci 105:309–318

Pal DK, Wani SP, Sahrawat KL, Srivastava P (2014) Red ferruginous soils of tropical Indian environments: a review of the pedogenic processes and its implications for edaphology. Catena 121:260–278. doi:10.1016/j.catena2014.05.023

Pal DK, Wani SP, Sahrawat KL (2015) Carbon sequestration in Indian soils: present status and the potential. Proc Natl Acad Sci Biol Sci (NASB) India 85:337–358. doi:10.1007/s40011-014-0351-6

Ray SK, Chandran P, Durge SL (2001) Soil taxonomic rationale: kaolinitic and mixed mineralogy classes of highly weathered ferruginous soils. Abstract, 66th Annual Convention and National Seminar on "Developments in Soil Science" of the Indian Society of Soil Science, Udaipur, Rajasthan, pp 243–244

Ray SK, Chandran P, Bhattacharyya T, Durge SL, Pal DK (2003) Layer charge of two benchmark vertisol clays by alkylammonium method. Clay Res 22:13–27

Reed MG, Scot AD (1962) Kinetics of potassium release from biotite and muscovite in sodium tetraphenylboron solutions. Soil Sci Soc Am Proc 26:437–440

Reichenbach HGV (1972) Exchange equilibria of interlayer cations in different particle size fractions of biotite and phlogopite. In: International clay conference proceedings, pp 457–466

Rich CI (1968a) Mineralogy of soil potassium. In: Kilmer et al. (eds) The role of potassium in agriculture. American Society of Agronomy, Madison, Wisconsin, pp 79–108

Rich CI (1968b) Hydroxy-interlayering in expansible layer silicates. Clays Clay Miner 16:15–30

Rich CI (1972) Potassium in soil minerals. In: Proceedings 9th colloquium, International Potash Institute, pp 3–12

Roy RN, Seetharaman S, Singh RN (1978) Soil and fertilizer potassium in crop nutrition—a review. Fertilizer News 23:3–26

Ryden JC, Pratt PF (1980) Phosphorus removal from wastewater applied to land. Hilgardia 48:1–36

Sahrawat KL (1995) Fixed ammonium and carbon–nitrogen ratios of some semi-arid tropical Indian soils. Geoderma 68:219–224

Sahrawat KL (2004a) Organic matter accumulation in submerged soils. Adv Agron 81:169–201

Sahrawat KL (2004b) Ammonium production in submerged soils and sediments: the role of reducible iron. Commun Soil Sci Plant Anal 35:399–411

Sahrawat KL, Warren GP (1989) Sorption of labeled phosphate by a Vertisol and an Alfisol of the semi-arid zone of India. Fertilizer Res 20:17–25

Sahrawat KL, Bhattacharyya T, Wani SP, Chandran P, Ray SK, Pal DK, Padmaja KV (2005) Long-term lowland rice and arable cropping effects on carbon and nitrogen status of some semi-arid tropical soils. Curr Sci 89:2159–2163

Sanyal SK, DeDatta SK (1991) Chemistry of phosphorus transformations in soil. In: Stewart BA (ed) Advances in soil science. Springer, New York, pp 1–120

Sarma VAK (1976) Mineralogy of soil potassium. Bull Indian Soc Soil Sci 10:66–77

Sarma VAK (1984) Mechanisms and rate of release of potassium from potassium bearing minerals in soils. Mineralogy of soil potassium. PRII Review Series I, Gurgaon, Haryana, pp 55–61

Shailaja S, Sahrawat KL (1994) Phosphate buffering capacity and supply parameters affecting phosphorus availability in Vertisols. J Indian Soc Soil Sci 42:329–330

Sharma KL, Mandal UC, Srinivas K, Vittal KPR, Mandal B, Grace JK, Ramesh V (2005) Long-term soil management effects on crop yields and soil quality in a dry land Alfisol. Soil Tillage Res 83:246–259

Smith GD (1986) The Guy Smith interviews: rationale for concept in soil taxonomy. SMSS Technical Monograph 11. SMSS, SCS, USDA, USA

Srinivasarao Ch, Venkateswarlu B, Lal R, Singh AK, Kundu S (2013) Sustainable management of soils of dryland ecosystems of India for enhancing agronomic productivity and sequestering carbon. Adv Agron 121:253–329

Srivastava P, Bhattacharyya T, Pal DK (2002) Significance of the formation of calcium carbonate minerals in the pedogenesis and management of cracking clay soils (Vertisols) of India. Clays Clay Miner 50:111–126

Talibudeen O, Weir AH (1972) Potassium reserves in a 'Harwell' series soil. J Soil Sci 23:456–474

Truc MT, Yoshida M (2011) Effect of zeolite on the decomposition of organic matter in tropical soils under global warming. World Acad Sci Eng Technol 59:1664–1668

Velayutham M, Pal DK, Bhattacharyya T, Srivastava P (2002) Soils of the Indo-Gangetic Plains, India–the historical perspective. In: Abrol YP, Sangwan S, Tiwari M (eds) Land use–historical perspectives–Focus on Indo-Gangetic Plains. Allied Publishers, New Delhi, pp 61–70

Warren GP, Sahrawat KL (1993) Assessment of fertilizer P residues in a calcareous Vertisol. Fertilizer Res 34:45–53

Wilson MJ (1999) The origin and formation of clay minerals in soils: past, present and future perspective. Clay Miner 34:7–25

Chapter 9
Importance of Pedology of Indian Tropical Soils in Their Edaphology

Abstract Many consider that the soils of the tropical soils are acidic, infertile, and that they do not support a reasonable sustained agricultural production. Recent research in agricultural production in tropical Asia and Latin America indicate that universal infertility of tropical soils is a myth. In India, during the green revolution period, a renaissance was initiated by the National Agricultural Research Systems (NARS) in a modest way by managing the tropical soils properly for their restoration and preservation. It is clear now that the substrate quality of Indian tropical soils is good enough to support the agricultural land uses, horticultural, spices and cash crops, in making India self-sufficient in food production. The substrate quality is maintained by progressive pedogenic processes (pedology) in tropical Indian soils, which are inherently linked to many edaphological issues. Recent advances in pedology of the Indian tropical soils have demonstrated their considerable potential and also amply established the basic necessity of pedological research, in better understanding some queer edaphological aspects of Indian tropical soils (Vertisols, RF soils and IGP soils), which are affected by the climate change during the Holocene period. Thus edaphology is inherently based on deep fundamental understandings of soils and thus basic pedological research in tropical soils needs to be encouraged vigorously to link some of their major unresolved edaphological aspects to develop improved management practices as guiding principles to improve and maintain soil health through adequate national recommended practices in other tropical parts of the world.

Keywords Indian tropical soils · Dissolution of myth · Soil substrate quality · Basic pedological research · Edaphology

9.1 Introduction

It is often believed that the soils of the tropics are acidic, infertile, and that they do not support a reasonable sustained agricultural production. However, recent research in agricultural production in tropical Asia and Latin America (FAO 1986)

© Springer International Publishing AG 2017
D.K. Pal, *A Treatise of Indian and Tropical Soils*,
DOI 10.1007/978-3-319-49439-5_9

indicate that universal infertility of tropical soils is a myth (Sanchez and Logan 1992). A renaissance is currently taking place in soil science (Hartemink and McBratney 2008). In India, during the green revolution period, such renaissance was initiated by the National Agricultural Research Systems, NARS in a modest way. This strategy commenced to manage the tropical soils properly for their restoration and preservation. It is thus important to highlight the substrate quality of Indian tropical soils, which have been efficiently supporting the agricultural land uses, horticultural, spices and cash crops, in making India self-sufficient in food production (Bhattacharyya et al. 2014a; Pal et al. 2012a, 2015). The substrate quality is maintained by progressive pedogenic processes (pedology) in tropical Indian soils, which are inherently linked to many edaphological issues. Recent advances in pedology of the Indian tropical soils (Pal et al. 2012a, 2014; Srivastava et al. 2015) have demonstrated their considerable potential and also amply established the basic necessity of pedological research especially in terms of pedogenetic processes, in better understanding some queer edaphological aspects of Indian tropical soils (Vertisols, RF soils and IGP soils), which are affected by the climate change during the Holocene period. It is thus important to highlight some selected edaphological issues where linking pedology and edaphology has helped to better understand the Indian tropical soils.

9.2 Impact of Spatially Associated Non-sodic (Aridic Haplusterts) and Sodic (Sodic Haplusterts) Vertisols on Crop Performance

Natural chemical soil degradation is common in SAT soils of India. In Vertisols of the Purna Valley of Maharashtra, central India (total area ∼0.6 M ha), natural degradation in terms of PC formation and development of subsoil sodicity is triggered by the semi-arid climate, with an MAR (mean annual rainfall) of 875 mm, a tropustic moisture regime and a hyperthermic temperature regime (Balpande et al. 1996; Pal et al. 2000a). Due to these pedogenetic processes, these soils have developed severe drainage problems, but in the Pedhi Watershed, in the adjacent east upland of the Purna valley (area ∼45,000 ha), Vertisols do not have any problem of subsoil sodicity but have drainage problems. The area, however, has a higher MAR (975 mm) than the Purna valley and has similar moisture and temperature regimes to the Purna Valley. Vertisols are the dominant soil type in the watershed, but as a result of micro-topographic variation (0.5–5 m; please refer to Fig. 2.6), Sodic Haplusterts occur on micro-high (MH) positions and at a distance of approximately 6 km, whereas Aridic Haplusterts occur on micro-low (ML) positions (Vaidya and Pal 2002). Following typical managements as detailed elsewhere (Kadu et al. 2003) farmers get better yield of cotton (0.6–1.6 t/ha of seed + lint) in Aridic Haplusterts with ESP <5 than in Aridic Haplusterts (ESP >5, <15; 0.6–1.0 t/ha) and Sodic Haplusterts (ESP ≥15; 0.2–0.8 t/ha yield;

Table 9.1 Range in values of PC, ESP, shC and yield of cotton in Vertisols of Vidarbha, Central India

District, Vidarbha Region, Maharashtra, Central India	Soil classification	PC (%)[a]	ESP[b]	shC[c] (mm h^{-1}) weighted mean in the profile (1 m)	Cotton yield (t ha^{-1}) (seed + lint)
Nagpur (MAR—1011 mm)	Typic Haplusterts/Typic calciusterts	3–6	0.5–11	4–18	0.9–1.8
Amravati (MAR—975 mm)	(a) Aridic Haplusterts	3–7	0.8–4	2–19	0.6–1.6
	(b) Sodic Haplusterts	3–13	16–24	0.6–9.0	0.2–0.8
Akola (MAR—877 mm)	(a) Aridic Haplusterts	3–4	7–14	3–4	0.6–1.0
	(b) Sodic Haplusterts	3–4	19–20	1–2	0.6

[a]*PC* pedogenic $CaCO_3$, [b]*ESP* exchangeable sodium percentage (sodicity), [c]*sHC* saturated hydraulic conductivity (Adapted from Kadu et al. 2003)

Table 9.1). Due to comparatively poor crop productivity of the Aridic Haplusterts with the Sodic Haplusterts with an ESP >5 but <15 (having no soil modifiers; Pal et al. 2006), Aridic Haplusterts were classified as Sodic Haplusterts (Balpande et al. 1996; Pal et al. 2006). Such a close association of Aridic and Sodic Haplusterts under similar topographical conditions in a relatively small watershed is a unique example in pedological parlance, but such occurrence of Vertisols poses a challenge for land resource managers in comprehending the differences in the chemical environment between the Aridic Haplusterts with ESP <5 and the Aridic Haplusterts with ESP >5 but <15. Thus, for optimized use and management of the latter type of Aridic Haplusterts, the proposed modifications in their subgroup-level classification (as per US Soil Taxonomy) are mandatory (Pal et al. 2006, 2012a).

9.3 Linear Distance of Cyclic Horizons in Vertisols and Its Relevance to Agronomic Practices

It is well known that Vertisols develop deep, wide shrinkage cracks in the summer, which close as the soil rewets. In the sub-surface regions where sphenoids and/or slickensides are formed, the difference between horizontal stress and vertical stress is quite large when swell. The cyclic horizons repeat in the subsoil, the size of which depends on the length of cycle. One-half of the linear distance (LD) of the cycle is a measurement of the lateral dimension of a cyclic horizon (Johnson 1963). For evaluating the subsoil variability and determine LD, trenches of at least 10 m long with depths of 2 m or more are required (Bhattacharjee et al. 1977; William et al. 1996). Such field examination is time consuming, laborious and expensive.

A large area of Vertisols is used for pastures, and cracks developed therein may be wide enough to cause dangerous footing for animals (Buol et al. 1978). Agronomic uses of Vertisols vary widely, depending on the climate. Field moisture conditions, drainage conditions and patterns of vegetation indicate that the maximum oscillation between wet and dry conditions manifests in micro-depressions that retain moisture for longer periods and in microknolls, which dry out faster. Vertisols are capable of tilting large trees, and surprisingly, few if any commercial forests are cultivated on Vertisols (Buol et al. 1978). In addition, highways, buildings, fences, pipelines, and utility lines are moved and distorted by the shrinking and swelling of these soils. Prior knowledge about the highs and lows of the cyclic pedons may help the stakeholders plan their programmes and avoid mishaps. In view of the need for a method to determine LD, a mathematical equation has been proposed to measure the LD of the cyclic horizons, taking into account the depth of occurrence of slickensides (Bhattacharyya et al. 1999). A standard parabolic equation represented by ($y^2 = 4ax$) was considered, where a is the focus of the parabola. The concept of cyclic horizons of Vertisols in terms of this parabolic pattern (Fig. 9.1) centres around two basic assumptions. The first assumption is that the depth of the first occurrence of slickensides (b) coincides with the focus of the parabola. The second assumption is that b remains constant within a cyclic horizon.

To calculate LD using the equation (LD = MN = 2KN = 4[a $(a + b)]^{1/2}$ (Fig. 9.1), the values of a and b are needed. In the field, the value of b

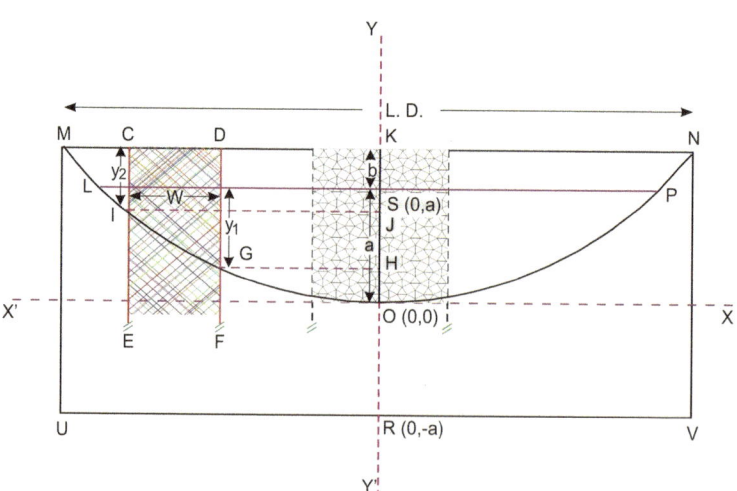

Fig. 9.1 The parabolic path of cyclic pedon where PL, S, UV, O, and KR are latus rectum, focus, directrix, vertex and axis of the parabola (KN = KM, OS = OR, KN = NV, KM = MU, OJ − OH = y_1 − y_2 = HJ, OJ + OH = y_1 + y_2, OJ. OH = y_1, y_2) (Adapted from Bhattacharyya et al. 1999)

(the depth of the first occurrence of slickensides) is obtained; however, the value of *a* can be obtained only by examining the profile exactly in the centre of the cyclic horizon. This value is difficult to acquire because from the surface the cyclic horizons in the subsurface cannot be identified, especially where micro-knolls and micro- depressions are obliterated. When the profile is examined away from the centre of the cyclic horizon, the value of *a* cannot be obtained, and calculation of LD becomes difficult. To circumvent this difficulty, Bhattacharyya et al. (1999) proposed the following equation.

LD (cm) = $200/Y (2500 + bY)^{1/2}$, where $Y = y_1 + y_2 - 2 (y_1 y_2)^{1/2}$ and y_1 and y_2 are the vertical distances (cm) from the first occurrence of slickensides to the intersecting points of the cyclic pedon such that $y_1 > y_2$ and *b* is the depth of the first occurrence of slickensides. It is always possible to find the values of y_1 and y_2, if the profile is examined on either side of the centre of the cyclic horizon (Fig. 9.1). This equation can eliminate some of the need for fieldwork to determine the LD with the help of three variables, namely y_1 (the length of the slickensided zone on the right-hand side of the profile wall from the depth of the occurrence of slick-ensides), y_2 (the length of the surface and the slickensided zone on the left-hand side of the profile wall) and *b* (the depth of the occurrence of slickensides). These values can be obtained with ease by soil survey and mapping. The accuracy of the equation proposed by Bhattacharyya et al. (1999) is between 81 and 86% in Vertisols in arid to semi-arid climates. The equation provides a new method of locating micro-depressions and micro-knolls in an effort to better manage Vertisols for agricultural and non-agricultural purposes.

9.4 Smectite as the First Weathered Product of the Deccan Basalt and Its Contribution in Growing Vegetation on Weathered Basalt and in Very Shallow Cracking Clay Soil

Dominated by dark ferromagnesian minerals, gabbros and basalts are more easily weathered than granites and other light-coloured rocks. Low charge dioctahedral smectite (DOS) is the first weathering product of the Deccan basalt. Therefore, the DOS rich alluvium of the weathering Deccan basalt is principally responsible for the formation of deep black soils in the Peninsular India (Pal and Deshpande 1987; Pal et al. 2009a). In the Deccan basalt litho logs, boulders of different sizes adjoin each other (Fig. 9.2a) and exhibit spheroidal weathering (Fig. 9.2b). These boulders have concentric rings similar to onions (Fig. 9.2c) that easily come away under gentle pressure from fingers. The boulders also contain DOS (Fig. 9.2d) similar to that of Vertisols (Pal and Deshpande 1987). Smectites do possess an excellent capacity to hold moisture and nutrients; thus, they help several tree species to anchor on weathered basalt (Fig. 9.2e), even on steep slopes (>40%). The long-term preservation of tree species under national forest management in Maharashtra and

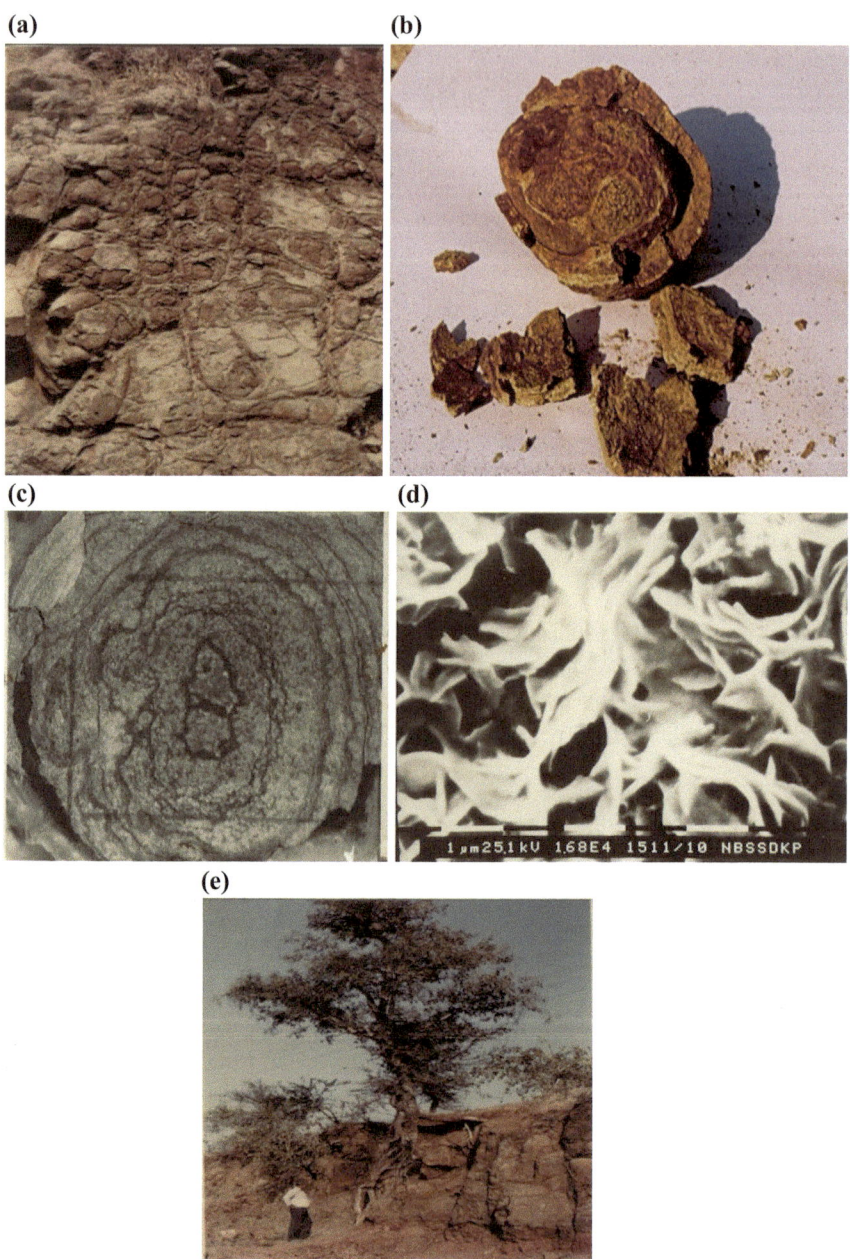

Fig. 9.2 Deccan basalt litholog showing spheroidal weathering (**a**), onion like peeling in basalt boulders (**b**), thin section of basalt boulder showing concentric weathering rinds (**c**), SEM *photograph* of clay sized dioctahedral smectite as its first weathering product (**d**) and smectite providing moisture and nutrient to tree species for anchoring on weathered basalt, even on steep slopes (>40%) (Adapted from Pal et al. 2000b and Pal and Deshpande 1987) *Photographs* (**a–e**), courtesy of DKP

Madhya Pradesh of western and central Peninsular India has become possible under favourable MAR conditions in HT, SHM, SHD, and SAM climates. In SAD and AD climates, care is needed at the initial stage of establishment of tree species. The largest Deccan basalt area with the greatest forest cover lies in the state of Madhya Pradesh, in central India; it is followed by Maharashtra, in western India. Smectite mineral has been very helpful in preserving and sustaining the spectacular natural forest vegetation (Bhattacharyya et al. 2005). The mineral also helps to maintain and preserve the forests in lower-MAR areas; however, initial care needed during the establishment of tree species. Thus, the natural abundance of smectite in the Deccan basalt areas of lower MAR can also prove useful in establishing agri-horticultural crops even on shallow soils (Entisols; depth <50 cm) strewn with small and weathered basalt rocks.

9.5   Sustainability of Rice Production in Zeolitic Vertisols

Vertisol use is not confined to a single production system. In India, major combinations of rainfed crops under semi-arid climatic environments are sorghum/pigeon pea, cotton/pigeon pea and cotton/sorghum/pigeon pea. Mixed cropping usually combines crops with different maturity lengths, drought-sensitive with drought-tolerant crops, cereals with legumes, and cash crops with food crops (Swindale 1989). In semi-arid western India (<1000 mm MAR), sugarcane and rice are grown under irrigated conditions, whereas rice is grown under rainfed conditions in areas of sub humid moist climatic conditions of central India (with MAR >1400 mm)(Pal et al. 2003a). The saturated hydraulic conductivity (sHC) decreases rapidly with depth in Vertisols, but the decrease is sharper in non zeolitic Sodic Haplusterts; and the weighted mean of sHC in 0–100 cm depth is <10 mm h^{-1}. In non-zeolitic Typic Haplusterts, the sHC is > 10 mm h^{-1}. But in zeolitic Typic Haplusterts (Kheri soils at Jabalpur, Tomar et al. 1996 and Sakka soils at Dindori, Madhya Pradesh, Pal et al. 2003a) sHC is >20 mm h^{-1}, and these soils are cultivated to rice as rainfed crop in areas with > 1400 mm MAR. Interestingly, zeolitic Sodic Haplusterts (Teligi soils in Bellary, Karnataka, Pal et al. 2013a) in areas with <700 mm MAR also have sHC >20 mm h^{-1}; and rice is cultivated in these soils under canal irrigation. The enhanced sHC (>20 mm h^{-1}) in such Vertisols due to the presence of zeolite, appears to be just adequate for the period of submergence required for the rice crop, and post-rainy season crops are successfully grown with good yields. Morphological examination of such soils showed no sign of gleyed horizons and soil moisture regime does not reach aquic conditions. Such situations are unique in nature and pose a great challenge to the soil mappers to classify them according to the US Soil Taxonomy, as they have good productive potential despite being sodic in nature. Sustainability of rice cropping system in such soils will, however, depend on rate of dissolution of Ca-zeolite on a timescale, and a new research initiative on this topic is warranted.

9.6   Holocene Climate Change, Natural Soil Degradation, Modified Vertisols and Their Evaluation

During the Quaternary Period, frequent climatic changes occurred (Ritter 1996). Due to climate change, soils worldwide were subjected to climatic fluctuations, especially in the last post-glacial period. Brunner (1970) reported evidence for tectonic movements during the Plio-Pleistocene transition, which caused the formation of various relief types. With the formation of the Western Ghats during the Plio-Pleistocene crustal movements, the humid climate of the Miocene-Pliocene was replaced by the semi-arid conditions that still prevail in central and southern peninsular India. The Arabian Sea flanks the Western Ghats, which rise precipitously to an average height of 1200 m, the result of a heavy orographic rainfall all along the west coast. The lee-side towards the coast receives less than 1000 mm of rainfall and is typically rain shadowed (Rajaguru and Korisetter 1987). The current aridic environment prevailing in many parts of the world (including India; Eswaran and van den Berg 1992) may create adverse physical and chemical soil environments, representing a regressive pedogenesis (Pal et al. 2013b). This is evident from the occurrence of more alkaline, calcareous and sodic shrink-swell soils (Sodic Haplusterts/Calciusterts) of Peninsular India due to a progressive increase in the PC content from HT to AD climates (Pal et al. 2009a; please refer to Fig. 2.12b), as aridity of the climate is the main factor in the formation of calcareous sodic soils (Pal et al. 2000a). The subsoils (SAM Aridic Haplusterts, SAD Sodic Haplusterts and AD Sodic Calciusterts) remain under less water than those of Typic Haplusterts in HT, SHM and SAM climates. As a result, the Vertisols in drier regions of India have relatively more PC and ESP, reduced sHC, (Please refer to Table 2.4) and poor micro-structure (please refer to Fig. 2.11c), as well as deep cracks cutting through the slickensided zones (please refer to Fig. 2.2). Thus, these modified Vertisols qualify as polygenetic soils (Pal et al. 2009a). Deep-rooted crops on Aridic Haplusterts (ESP >5, <15) and Sodic Haplusterts show poor productivity (Table 9.1). Recently, Kadu et al. (2003) and Deshmukh et al. (2014) attempted to identify bio-physical factors that limit the yield of deep-rooted crops (cotton) in 32 Vertisols (developed in the basaltic-alluvium) of the Nagpur, Amravati and Akola districts in the Vidarbha region in central India. Under rain-fed conditions, the yield of deep-rooted crops on Vertisols depends primarily on the amount of rain stored at depth in the soil profile and the extent to which this soil water is released between the rains during crop growth. Both the retention and release of soil water are governed by the nature and content of clay minerals, and also by the nature of the exchangeable cations. The AWC (available water content), calculated based on moisture content, varied between 33 and 1500 kPa (Table 9.2), indicating that not only the Typic/Aridic Haplusterts but also the Sodic Calciusterts can hold sufficient water; however, a non-significant negative correlation between cotton yield and AWC (Table 9.2) indicates that this water is not available during the growth of crops. The Na^+ ions on exchange sites of Aridic Haplusterts and Sodic Haplusterts/Calciusterts with ESP >5 thus amounts to over estimation

(Gardner et al. 1984). In fact, moisture remains at 100 kPa for Typic Haplusterts and Aridic Haplusterts (ESP <5) after the cessation of rains during June to September, while it is held at 300 kPa for Sodic Haplusterts (Pal et al. 2012a; Deshmukh et al. 2014) as the movement of water is governed by sHC, which decreases rapidly with depth, and the decrease is sharper in Aridic/Sodic Haplusterts (ESP >5, Pal et al. 2009a). This conclusion is supported by a significant positive relationship between ESP and AWC, and a significant negative correlation between yield and ESP (Table 9.2). A significant positive correlation between yield and exchangeable Ca/Mg (Table 9.2) indicates that a dominance of Ca^{2+} ions in the exchange sites of Vertisols is required to improve the hydraulic properties for a favourable growth and final yield of crops. The development of subsoil sodicity (ESP ≥5) replaces Ca^{2+}ions in the exchange complex, causing a reduction in the yield of cotton in Aridic/Sodic Haplusterts (ESP ≥5). A significant negative correlation between ESP and exchangeable Ca/Mg (Table 9.2) indicates an impoverishment of soils with Ca^{2+} ions during sodification by the illuviation of Na-rich clays. This pedogenetic process depletes Ca^{2+}ions from the soil solution in the form of $CaCO_3$, with the concomitant increase of ESP with pedon depth. Thus, these soils contain PC (Pal et al. 2000a), and carbonate clay, which, on a fine earth basis, increases with depth (please refer to Table 2.4). This chemical process is evident from the positive correlation between ESP and carbonate clay (Table 9.2). A significant positive correlation between the yield of cotton and carbonate clay (Table 9.2) indicates that, like ESP, PC formation also causes a yield reduction and is a more important soil parameter than total soil $CaCO_3$ (NBSS&LUP 1994;

Table 9.2 Co-efficient of correlation among various soil attributes and yield of cotton[a]

No.	Parameter Y	Parameter X	r
Based on 165 soil horizons samples of 29 Vertisols			
1	sHC (mm h^{-1})	Exch. Ca/Mg	0.51*
2	sHC (mm h^{-1})	ESP[b]	−0.56*
3	ESP	AWC (%)	0.40*
4	ESP	Exch. Ca/Mg	−0.40*
Based on 29 Vertisols			
5	Yield of cotton (q ha^{-1})	AWC (%) WM[c]	−0.10
6	Yield of cotton (q ha^{-1})	ESP max[a]	−0.74*
7	Yield of cotton (q ha^{-1})	sHC WM[b]	0.76*
8	Yield of cotton (q ha^{-1})	Carbonate clay[d]	−0.64*
10	Yield of cotton (q ha^{-1})	Exch. Ca/Mg WM[c]	0.50*
11	ESP max[a]	AWC (%) WM[c]	0.30*
12	ESP max[a]	Exch. Ca/Mg WM[c]	−0.55*
13	ESP max[a]	Carbonate clay[d]	0.83*

AWC available water content; *ESP* exchangeable sodium percentage; *sHC*, saturated hydraulic conductivity
*Significant at 1% level
[a]Adapted from Kadu et al. (2003); [b]Maximum in pedon;
[c] weighted mean; [d] fine earth basis

Sys et al. 1993). An accelerated rate of PC formation in dry climates impairs the hydraulic properties of Vertisols, and a significant negative correlation exists between ESP and sHC (Table 9.2). The processes operating in the soils of dry climates also influence the sHC of the Vertisols. A significant positive correlation exists between the yield of cotton and sHC. In view of the pedogenic relationship among SAT environments, PC formation, exchangeable Ca/Mg, ESP and sHC, all of which ultimately impair the drainage of Vertisols, the evaluation of Vertisols for deep-rooted crops based on sHC alone may help in planning and management of soils, not only of Vertisols in the Indian SAT areas, but also of Vertisols under similar climatic conditions elsewhere (Kadu et al. 2003; Pal et al. 2012a).

9.7 Soil Modifiers (Zeolites, Gypsum and CaCO₃) in Mitigating the Adverse Holocene Climate Change and Making Sodic Vertisols Resilient

Since the beginning of the 1990 the presence of soil modifiers such as gypsum, calcium carbonate and zeolites, is being adequately reported in soils by the researchers of the ICAR-NBSS & LUP, Nagpur (Pal 2013). It is necessary to follow the unique role of soil modifiers in changing the pedo-chemical environment of soils, which is essential for better soil use and management, in fine-tuning the exiting soil classification scheme and also the management practices to enhance and sustain the productivity of Indian tropical soils (Pal 2013).

9.7.1 Ca-Zeolites

During the last 3 decades, zeolite minerals have been recognized with increasing regularity as common constituents of Cenozoic volcanogenic sedimentary rocks and altered pyroclastic rocks (Ming and Mumpton 1989). Zeolites have also been reported as secondary minerals in the Deccan flood basalt of the Western Ghats in the state of Maharashtra, India (Sabale and Vishwakarma 1996). Among the commonly occurring species of zeolites, heulandites (Fig. 9.3a) have a wide occurrence both in time and space (Sabale and Vishwakarma 1996). Zeolites have the ability to hydrate and dehydrate reversibly and to exchange some of their constituent cations. Consequently, they can influence the pedochemical environment during the formation of soils. The significance of zeolites has recently been realized in the formation and persistence of slightly acidic to acidic Vertisols (Typic Haplusterts) in HT climatic environments, not only in central and western India (Bhattacharyya et al. 2005), but elsewhere (Ahmad 1983). The formation and persistence of Vertisols on the Deccan basalts in HT climate with other associated soils (Mollisols and Alfisols) due to the presence of Ca-zeolites is a unique example

(a)   **(b)**

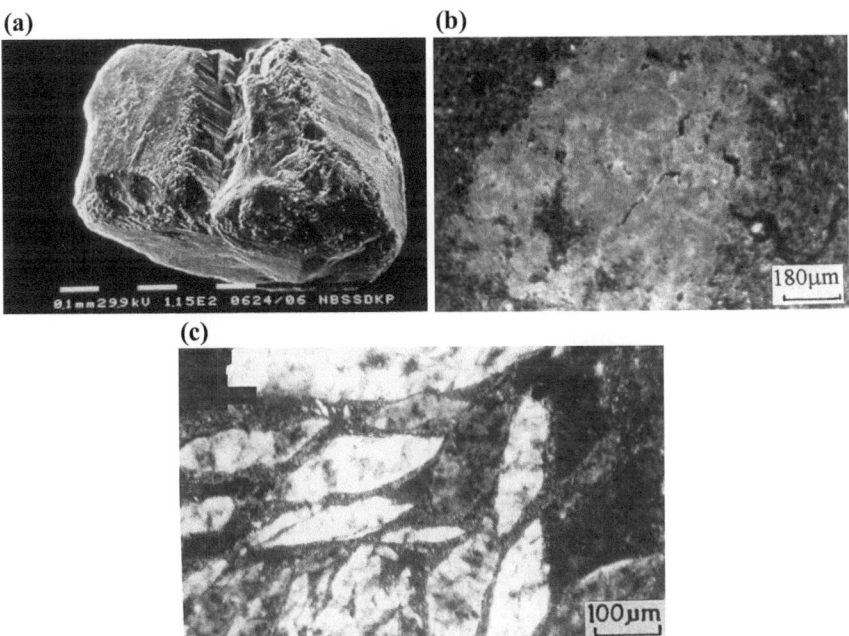

(c)

Fig. 9.3 Scanning electron microscope photographs of **a** unweathered heulandite of shrink-swell soils, and photomicrographs of **b** pedogenic calcium carbonate of shrink-swell soils, **c** lenticular gypsum crystal of shrink-swell soils (Adapted from Pal et al. 2000b)

in an open system such as soil. This highlights the inadequacy of the existing models to explain the formation of Indian tropical soils and also the role of zeolites in preventing the loss of soil productivity and maintaining soil health (as elaborated in Chap. 5).

Many productive Vertisols under rain-fed conditions have been rendered unproductive for agriculture under irrigated conditions in the longer-term. However, some zeolitic Vertisols of the SAD parts of western India have been irrigated through canals for the last twenty years to produce sugarcane. These soils lack salt-efflorescence on the surface and are not waterlogged at present, suggesting that these soils are not degraded due to their better drainage. However, these soils are now Sodic Haplusterts in view of their pH, ECe and ESP values, but they have sHC >10 mm h^{-1} (weighted mean in the 0–100 cm, Table 9.3). A constant supply of Ca^{2+} ions from Ca-zeolites in these soils most likely helps maintain a better drainage system. Because of such natural endowment with a soil modifier, no ill effects of high ESP (>15) in crop production in the Vertisols of Gezira in Sudan (El Abedine et al. 1969; Robinson 1971) and in Tanzania (Ahmad 1996) were observed. In addition, some Vertisols of the AD climate of western India produce deeply rooted crops such as cotton under rain fed conditions comparable to those of the Typic Haplusterts of the SAM climate of central India (Pal et al. 2009b). The sHC (weighted mean, 0–100 cm) of these soils is >10 mm h^{-1} (Table 9.3), despite

Table 9.3 Adverse effect of irrigation on zeolitic Typic Haplusterts during sugarcane cultivation

Depth (cm)	pH (1:2)	ECe (dS m^{-1})	CaCO$_3$ (<2 mm) %	ESP	sHC[a] (mm h^{-1})	Base saturation (%)	Available K (kg ha^{-1})
0–20	9.0	0.77	16.0	4.2	18	107	686
20–42	9.2	1.01	17.0	10.4	17	119	343
42–68	9.3	0.99	17.0	18.8	5	94	343
68–102	9.0	1.25	15.0	13.7	10	105	343
102–131	9.0	1.09	25.3	12.1	13	103	343
131–150	9.0	1.02	16.1	8.0	12	109	421

Adapted from Pal et al. (2003a)
ECe electrical conductivity of the saturation extract; *ESP* exchangeable sodium percentage; *sHC* saturated hydraulic conductivity
[a] 13 mm h^{-1} is the weighted mean sHC in 0–100 cm depth of soil

being Sodic Calciusterts (Pal et al. 2009a). However, the sustainability of crop productivity in the dry climate depends on the solubility and supply of Ca^{2+}ions from zeolites such that it is sufficient to overcome the ill effects of the pedogenic threshold of dry climates (Pal et al. 2003b, 2009a). Such situations are unique in nature and pose a great challenge to soil mappers to classify them as per the US Soil Taxonomy when they are sodic in nature but have good productive potential.

9.7.2 Gypsum

Arid and semi-arid environments trigger natural soil degradation processes in terms of the precipitation of CaCO$_3$ and the concomitant development of sodicity (Pal et al. 2000a). Despite this possibility, selected Vertisols of the SAD climate in southern India are non-sodic and support the growth of crops such as cotton, pigeon pea and sorghum. The development of sodicity has been prevented by the presence of gypsum in these soils (Fig. 9.3b), but the soils are calcareous in nature (Table 9.4). The soils have a sHC >30 mm h^{-1} (Table 9.4) despite the rapid formation of PC, unlike in the zeolitic Vertisols of the SAD climate (Pal et al. 2009a). This can be attributed to the greater solubility of gypsum (30 me/L) than that of Ca-zeolites (< 0.1 me/L) in distilled water (Pal et al. 2006). The gypsum in such soils is antagonistic to the formation of more soluble salts in soils, as it prevents clay dispersion. Although the sustainability of crop productivity in these soils depends on the gypsum stock, the present poor productivity of cotton (approximately 2 t ha^{-1}) may be enhanced by irrigation because the gypsum present would prevent water logging and maintain better drainage (Pal et al. 2009c).

Table 9.4 Physical and chemical properties of Vertisols modified by gypsum in SAT parts of Tamil Nadu

Depth (cm)	pH (1:2 water)	ECe (dS/m)	CaCO₃ (<2 mm) (%)	ESP	sHC (mm/h)
0–6	8.0	0.2	5.4	0.5	19
6–20	8.0	0.3	4.3	0.9	22
20–41	8.0	0.5	5.3	0.6	44
41–74	8.0	0.4	7.9	0.9	30
74–104	7.9	0.2	12.5	1.1	37
104–128	7.9	0.6	12.8	1.4	34
128–140	7.4	2.7	15.6	1.8	32
140+	7.5	–	17.4	0.3	48

Adapted from Pal et al. (2003a). *ESP* exchangeable sodium percentage, *ECe* electrical conductivity of the saturation extract, *sHC* saturated hydraulic conductivity

9.7.3 CaCO₃

The subsoil sodicity impairs the hydraulic properties of the Vertisols of SAT environments, and this leads to the formation of sodic soils with ESP decreasing with depth. These soils are impoverished in organic carbon but have become enriched with $CaCO_3$ (Fig. 9.3c) with poor sHC (Table 9.5) (Pal et al. 2009a). However, such soils show enough resilience under the improved management (IM) (catchment management followed by adopting legume-based crop rotation, improved nutrient management and without any chemical amendments) system of ICRISAT, implemented in Patancheru, India. Through the implementation of such practices, a substantial increase in soil organic carbon (SOC) stock was observed (Wani et al. 2003). The resilience of such soils has been maintained by implementing IM practices in Vertisols. The increase in SOC is, however, related to chemical changes after the specified management interventions. In Vertisols (Sodic Haplusterts, Pal et al. 2012c) after 30 years of IM, the weighted mean (WM) of sHC in the first 100 cm of the profile increased by almost 2.5 times due to the reduction of ESP through the dissolution of $CaCO_3$ (Table 9.5), making the soils more permeable to air and water. In the last 24 years (since 1977), the rate of dissolution of $CaCO_3$ has been 21 mg year^{-1} in the first 100 cm of the profile. Dissolved Ca^{2+} ions improve the Ca/Mg ratio on the exchange complex of soils under IM compared with those under traditional management (Table 9.5) (Pal et al. 2012b).

The changes in soil properties, as stated above, suggest that $CaCO_3$ is dissolved through the cations of acidic root exudates and carbonic acid that formed due to evolved CO_2 from the root respiration in an aqueous solution, resulting in the formation of Ca $(HCO_3)_2$. The soluble Ca $(HCO_3)_2$, therefore, helps restore both the soluble and exchangeable Ca ions in the soils. The ESP decreases and the soil structure improve; as a result, the hydraulic properties of soils are improved (Table 9.5). This improvement in soil properties highlights the role of $CaCO_3$,

Table 9.5 Modification of physical and chemical properties of Vertisols through improved management system at ICRISAT, Patancheru in 24 years since 1977

Horizon	Depth cm	Clay (%), weighted mean in 0–100 cm	Clay %	Fine clay (%), weighted mean in 0–100 cm	Fine clay %	sHC mm h⁻¹, weighted mean in 0–100 cm	sHC mm h⁻¹	pH H₂O (1:2)	Organic carbon %	Organic carbon (%) weighted mean in 0–100 cm	CaCO₃ %	CaCO₃ (%), weighted mean in 0–100 cm	CEC cmol (p+) kg⁻¹	CEC (cmol(p+) kg⁻¹), weighted mean in 0–100 cm	Exchangeable Ca/Mg	Exchangeable Ca/Mg, weighted mean in 0–100 cm	ESP	ESP weighted mean in 0–100 cm
Kasireddipalli soil (Sodic Haplusterts) under traditional management (TM)																		
Ap	0–12	53.0	48.0	33.0	26.4	4.0	7.0	7.8	0.6	0.42	6.0	6.2	48.7	52.2	3.2	2.2	2.0	8.3
Bw1	12–30		51.4		29.7		6.0	7.8	0.4		6.2		52.1		2.8		4.0	
Bss1	30–59		52.5		32.5		6.0	8.1	0.4		6.0		52.2		2.1		7.1	
Bss2	59–101		55.6		36.4		2.0	8.5	0.4		6.4		53.5		1.8		13.0	
Bss3	101–130		59.4		30.8		2.0	8.5	0.4		6.5		57.8		3.1		8.0	
BCk	130–160		58.0		38.7		1.0	8.2	0.1		9.1		49.5		1.5		22.2	
Kasireddipalli soil (Typic Haplusterts) under improved management (IM)																		
Ap	0–12	54.7	52.1	32.8	28.8	11.0	17.0	7.5	1.0	0.53	4.2	5.7	50.4	56.0	2.9	2.4	2.0	4.5
Bw1	12–31		51.5		28.1		16.0	7.8	0.6		4.5		54.3		2.4		2.0	
Bss1	31–54		54.2		34.0		10.0	7.8	0.4		6.2		55.6		1.7		3.0	
Bss2	54–84		57.3		40.0		9.0	8.2	0.4		5.1		56.4		1.9		7.0	
Bss3	84–118		56.5		26.0		7.0	8.1	0.5		8.6		61.6		3.8		7.0	
Bss4	118–146		59.3		31.7		3.0	8.2	0.5		8.4		58.2		2.1		7.0	
BC	146–157		60.0		41.5		–	8.2	0.3		7.4		55.2		1.1		9.0	

Adapted from Pal et al. (2003a, 2012a, b, c)

which remains chemically inert (Pal et al. 2000a) during its sequestration (Sahrawat 2003) but acts as a soil modifier during the amelioration of degraded soils. The improvements in soil properties are also reflected in the classification of Vertisols. The original Kasireddipalli soils (Sodic Haplusterts) now qualify as Typic Haplusterts (Table 9.5) (Pal et al. 2012b).

SAT induced naturally degraded Vertisols of the Holocene period (with ESP >5, but <15 and sHC <10 mm h^{-1}) (Table 9.5) like Kasireddipalli and similar soils (Sodic Haplusterts, without soil modifiers like Ca zeolites and gypsum) occurring elsewhere show poor crop productivity. The IM system of ICRISAT when adapted in such soils can make Sodic Haplusterts resilient by converting them to Typic Haplusterts. In view of constant supply of soluble Ca^{2+} ions through the dissolution of $CaCO_3$, sustainability of still calcareous Typic Haplusterts can be maintained for couple of centuries under SAT environments. Therefore, the IM system can be considered as a good/recommended/no regrets strategy as it has potential to mitigate the adverse effect of climate change. It may thus possibly lessen the emphasis of genetically modified crops for Sodic Haplusterts of the SAT, and is ready for its wide adaptation through national and international initiatives (Pal et al. 2012b).

9.8 Acidity, Al Toxicity and Lime Requirement in RF Soils of HT Climates: A Critique

Indeed, soil acidity is a major constraint to crop production because of the significant contribution of Al_3^+ ions to exchangeable acidity in HT soils. To ward off the adverse effect of acidity, liming is generally recommended for strong to moderately acid soils of tropical India. But crop response to liming is not observed (Kadrekar 1979) in some moderately acidic Alfisols of the Western Ghats due to the presence of Ca-zeolites in these soils (Pal et al. 2012a). Aluminium toxicity to plants is a constraint in Oxisols, Ultisols and Dystropepts; and this is also associated with an overall low nutrient reserves (Sanchez and Logan 1992). Ultisols and Dystropepts of Kerala, Goa, Tamil Nadu, Karnataka and NEH are strongly to moderately acidic, and liming is often recommended to correct soil acidity and improve nutrient availability (ICAR-NAAS 2010).

Liming is done in order to bring exchangeable aluminium level in the soil to <1 mg kg^{-1} (Sanchez 1976) for maintaining sustained productivity because it improves the general nutrient availability (Nayak et al. 1996a, b; Sen et al. 1997). It is to be noted that reports on Al toxicity to crops in Indian acidic soils are scarce. The KCl extractable aluminium is used for liming tropical acidic pH soils (Oates and Kamprath 1983). It is intriguing that despite equally strongly acidic, Ultisols and Dystropepts of NEH, Ultisols of Kerala, and Alfisols of Goa indicate varying KCl exchange acidity. It is less than 1 cmol (p+)kg^{-1} for the Ultisols of Kerala and Meghalaya, and Alfisols of Goa whereas in the Ultisols of Arunachal Pradesh, Assam, Mizoram, Nagaland, Tripura and Dystrochrepts of Manipur, it ranges from 3 to 6 cmol (p+) kg^{-1} (please refer to Table 3.2). Thus, lime requirement (LR) of these acidic soils varies widely. Lime requirement is around 1 t lime ha^{-1} for Ultisols of Meghalaya, and for Dystrochrepts and Ultisols of other states of NEH, it ranges from 4 to 12 t ha^{-1}(please refer to Table 3.2). Such a low LR (<1 t ha^{-1}) may be expected for zeolitic soils as soil solution would remain adequately high in soluble Ca-ions from dissolution of Ca-zeolite but for the gibbsite containing Ultisols (Kerala, Meghalaya) and Alfisols (Goa), the reason for low KCl exchangeable acidity and LR, it is not understood yet. Therefore, fresh research on the role of zeolites and gibbsite in modifying the LR of acid soils needs to be initiated to make recommendations for LR on the basis of KCl exchangeable acidity. It is interesting to note that highly acidic pH Ultisols of Kerala, Karnataka, Tamil Nadu and NEH and moderately acidic Alfisols of Goa, Karnataka and Tamil Nadu have very high $BaCl_2$-TEA extractable acidity in contrast to that they have low to very low 1N KCl exchangeable acidity (please refer to Table 3.2). This indicates that Al_3^+ ions released during the humid tropical weathering do not remain in soil solutions but are trapped as Al $(OH)_2^+$ ion in the interlayers of 2:1 minerals to form hydroxy inter-layered minerals (Bhattacharyya et al. 2006). Both smectites and vermiculites act as sinks for aluminium and thus protect the biota from Al toxicity. It is evidenced when surface horizons (0–30 cm) of such Ultisols release large quantities of Al_3^+, Fe_3^+ and H^+ on extraction with $BaCl_2$-TEA (Pal et al. 2014). In the event of their availability in soil solutions these cations might have created very high acidity as well as Al-toxicity in soils to render them problematic for agricultural use.

9.9 Present Soil Health Status Due to Anthropogenic Activities in IGP Soils

It is crucial to understand the impact of climatic transition to aridification during 4th and 3rd millennium BC on cultivation over the IGP. Prior to the transition, rainfall in the Indus Valley was probably more than double the amount received now. Such a high rainfall facilitated the flourishing of both agriculture and forestry (Randhawa 1945). The aridification set in these times possibly caused the end of the great Harrapan civilization. The onset of climatic aridity started inducing regressive pedogenetic processes in the IGP soils, which made soils calcareous and sodic and also impaired their percolative moisture regime (Pal et al. 2000a). At present, the IGP covers approximately 1/3 cultivable area of India and produces nearly 50% of the country's food grains for its population. Most of the land in the IGP has been cultivated using traditional mixed cropping methods until the middle of the 19th century. During the last 3–4 decades the agriculturists of the IGP have successfully increased food grain production by introducing high-input technologies to meet the demands of the exponentially growing population. The strategies and measures adopted to achieve this success included, among others, (i) the spread of high-yielding varieties, (ii) expansion of irrigated area, (iii) increased use of fertilizers, (iv) use of plant protection chemicals, (v) strengthening of marketing infrastructure, and (vi) introduction of subsidies (Abrol et al. 2002). The production of grains was, however, not uniform across the IGP regions because of the spatial variation in land-resource characteristics and socio-economy in the region. These management interventions for 'money economy' have resulted in (i) widespread degradation, (ii) depletion of natural resources, (iii) declining water level, (iv) loss in soil fertility, (v) nutrient imbalance/deficiency, (vi) drainage congestion and (vii) loss in soil carbon (Abrol and Gupta 1998; Bhandari et al. 2002; Gupta 2003). Thus the sustainability ratings of some soil series of the IGP for the rice-wheat cropping system indicate many soil constraints, including low soil organic carbon (SOC) (Bhattacharyya et al. 2004).

The ICAR-NBSS&LUP reassessed the SOC stock of the IGP in 2005, and observed that there has been more than 30% increase in SOC in 0–150 cm depth over the value assessed in 1980. The SIC stock increased in some soils but decreased in sodic soils after their reclamation (Table 9.6) (Pal et al. 2015). The observed increase in SIC stock in the wetter part ((Table 9.6) could possibly be due to the accumulation of carbonates and bicarbonates from tube well water used for irrigation in the dry season (Pal et al. 2010). Despite the fact that the IGP soils of hot arid (HA), semi-arid dry (SAD), semi-arid moist (SAM), and sub-humid dry (SHD) climates may show deficiency in organic carbon due to high rate of decomposition, such decline in soil organic carbon did happen when the National Agricultural Research Systems' (NARS) recommendations for improved seeds, NPK fertilizers, micronutrients, FYM, and the inclusion of legumes in cropping sequence was not implemented in farmers' fields (Pal et al. 2015). This fact becomes clear when Indian tropical soils under various agricultural land uses under

Table 9.6 Changes in carbon stock (Tg = 10^{12} g) over years in the selected benchmark spots of the IGP (0–150 cm)

Bioclimatic systems	Soil series	SOC stock (Tg/10^5 ha)			SIC stock (Tg/10^5 ha)		
		1980	2005	SOC change over 1980 (%)	1980	2005	SIC change over 1980 (%)
Semi arid	Phaguwala	3.66	5.48	68	13.10	26.14	9
	Ghabdan	2.63	7.04	167	18.95	7.71	−59
	Zarifa	4.13	5.38	30	22.36	16.98	−24
	Viran	4.13	5.50	395	0	58.13	100
	Fatehpur	1.11	8.55	111	51.03	5.37	−89
	Sakit	4.05	5.84	31	0	10.15	100
	Dhadde	4.47					
Sub-humid	Bhanra	1.81	5.34	197	0	0.58	100
	Jagjitpur	2.52	8.76	248	2.52	8.86	251
	Haldi	8.55	6.28	−26	0	2.84	100
Humid	Hanrgram	6.93	11.0	59	0	3.68	100
	Madhpur	3.99	4.97	25	4.03	15.98	296
	Sasanga	5.25	8.42	61	0.88	4.45	405

Adapted from Bhattacharyya et al. (2007)

both short and long–long term experiments showed their potentiality to sequester OC under both arable and submerged conditions and they still show potential to sequester OC even in humid climates (Pal et al. 2015).

It is interesting to note that when the management interventions of the NARS have resulted in depletion of soil organic carbon in erstwhile Mollisols (e.g. Haldi soils, unit 26, Fig. 4.1; Table 9.6), the SIC stocks (as $CaCO_3$) in the arid and semi-arid IGP soils seem to be useful during the establishment of vegetation effected by appropriate ameliorative methods. During amelioration, the plant roots dissolve the immobile $CaCO_3$ and ultimately trigger the process of Ca release in the soil and thus $CaCO_3$ acts as a natural ameliorant for sodic soils (Bhattacharyya et al. 2004). It was observed that the rate of dissolution of $CaCO_3$ in Typic Natrustalfs was much higher than its rate of formation (details are available in Chap. 6) (Pal et al. 2000a, 2009c). This is important that $CaCO_3$ as SIC is more than double of the SOC stock in the first 150 cm depth, which helps improve the drainage, establishment of vegetation, and sequestration of OC in soils (Bhattacharyya et al. 2004) and would not allow Haplustalfs to transform to any other soil order so long $CaCO_3$ continues to serve as a soil modifier (Pal et al. 2012b).

In addition to the impairment of drainage by sodicity, some of the non-sodic soils are also marked by low hydraulic conductivity (sHC <10 mm/h) (Pal 2012). Impairment of hydraulic conductivity in such soils is attributed to an increase in bulk density (BD) in the subsoils (Pal 2012; Bhattacharyya et al. 2014b; Chandran et al. 2014; Tiwary et al. 2014). The increase in BD in subsoils is due to compaction caused by modern agricultural implements used to meet the high demand of rice, wheat and potato crops. This situation however, may help to maintain the yield of rice in rainy season and also in sequestering more soil organic carbon under sub-merged condition (Sahrawat 2004; Sahrawat et al. 2005). However, the yield of the subsequent crop, such as wheat, is either plateauing or declining (Dhillon et al. 2010). A comparison of the 1980 and 2005 soil data indicate an overall increase in SOC stock in these soils under agricultural land use despite the increase in soil inorganic carbon (SIC) (Bhattacharyya et al. 2007; Pal et al. 2009d). The soil degradation in terms of development of sodicity, increase in SIC and BD, despite the positive balance of OC sequestration, is a matter of concern. In view of the vast area of the IGP, new research initiatives are needed for development of the historical soil-climate-crop databank. Such a databank will not only help in fine-tuning the existing NARS management interventions, but also make the future projections on the sustainability of the cropping system in the IGP (Dhillon et al. 2010).

References

Abrol IP, Gupta RK (1998) Indo-Gangetic plains: issues of changing land use. LUCC Newsletter, March 1998, No. 3

Abrol YP, Sangwan S, Dadhwal VK, Tiwari M (2002) Land use/land cover in Indo-Gangetic Plains—history of changes, present concerns and future approaches. In: Abrol YP, Sangwan S,

Tiwari M (eds) Land use-historical perspectives—focus on Indo-Gangetic Plains. Allied Publishers Pvt. Ltd., New Delhi, pp 1–28

Ahmad N (1983) Vertisols. Pedogenesis and soil taxonomy. In: Wilding P, Smeck NE, Hall GF (eds) The soil orders, vol II. Elsevier, Amsterdam, pp 91–123

Ahmad N (1996) Occurrence and distribution of Vertisols. In: Ahmad N, Mermut AR (eds) Vertisols and technologies for their management. Elsevier, Amsterdam, pp 1–41

Balpande SS, Deshpande SB, Pal DK (1996) Factors and processes of soil degradation in Vertisols of the Purna valley, Maharashtra, India. Land Degrad Dev 7:313–324

Bhandari AL, Ladha JK, Pathak H, Padre A, Dawe D, Gupta RK (2002) Yield and soil nutrient changes in a long-term rice-wheat rotation in India. Soil Sci Soc Am J 66:162–170

Bhattacharyya T, Pal DK, Velayutham M (1999) A mathematical equation to calculate linear distance of cyclic horizons in Vertisols. Soil Surv Horiz 40:109–134

Bhattacharyya T, Pal DK, Chandran P, Mandal C, Ray SK, Gupta RK, Gajbhiye KS (2004) Managing soil carbon stocks in the Indo-Gangetic Plains, India. Rice–wheat consortium for the Indo-Gangetic Plains, New Delhi, p 44. (http://www.rwc-prism.cgiar.org and http://www.cimmyt.org)

Bhattacharyya T, Pal DK, Chandran P, Ray SK (2005) Land-use, clay mineral type and organic carbon content in two Mollisols–Alfisols–Vertisols catenary—sequences of tropical India. Clay Res 24:105–122

Bhattacharyya T, Pal DK, Velayutham M, Vaidya P (2006) Sequestration of aluminium by vermiculites in LAC soils of Tripura. Abstract, 71st Annual Convention and National Seminar on "Developments of Soil Science" of the Indian Society of Soil Science, Bhubaneswar, Orissa, p 1

Bhattacharyya T, Chandran P, Ray SK, Pal DK, Venugopalan MV, Mandal C, Wani SP (2007) Changes in levels of carbon in soils over years of two important food production zones of India. Curr Sci 93:1854–1863

Bhattacharyya T, Chandran P, Ray SK, Mandal C, Tiwary P, Pal DK, Wani SP, Sahrawat KL (2014a) Processes determining the sequestration and maintenance of carbon in soils: a synthesis of research from tropical India. Soil Horizons, p 1–16. doi:10.2136/sh14-01-0001 (Published 9 July 2014)

Bhattacharyya T, Sarkar D, Ray SK, Chandran P, Pal DK et al (2014b) Georeferenced soil information system: assessment of database. Curr Sci 107:1400–1419

Bhattacharjee JC, Landey RJ, Kalbande AR (1977) A new approach in the study of Vertisol morphology. J Indian Soc Soil Sci 25:221–232

Brunner H (1970) Pleistozäne Klimaschwankungen im Bereich den ostlichen Mysore Plateaus (Sudindien). Geologie 19:72–82

Buol SW, Hole FD, McCracken RJ (1978) Soil genesis and classification. Oxford and IBH Publ. Co., New Delhi

Chandran P, Tiwary P, Mandal C, Prasad J, Ray SK, Sarkar D, Pal DK et al (2014) Development of soil and terrain digital database for major food-growing regions of India for resource planning. Curr Sci 107:1420–1430

Deshmukh HV, Chandran P, Pal DK, Ray SK, Bhattacharyya T, Potdar SS (2014) A pragmatic method to estimate plant available water capacity (PAWC) of rainfed cracking clay soils (Vertisols) of Maharashtra, Central India. Clay Res 33:1–14

Dhillon BS, Kataria P, Dhillon PK (2010) National food security vis-à-vis sustainability of agriculture in high crop productivity regions. Curr Sci 98:33–36

El Abedine AZ, Robinson GH, Tyego V (1969) A study of certain physical properties of a Vertisol in a Gezira area, Republic of Sudan. Soil Sci 108:358–366

Eswaran H, van den Berg E (1992) Impact of building of atmospheric CO_2 on length of growing season in the Indian sub-continent. Pedologie 42:289–296

FAO (Food and Agriculture Organization of the United Nations) (1986) Yearbook of agriculture for 1985. FAO, Rome

Gardner EA, Shaw RJ, Smith GD, Coughlan KG (1984) Plant available water capacity: concept, measurement, prediction. In: McGarity JW, Hoult EH, Co HB (eds) Properties, and utilization of cracking clay soils. University of New England, Armidale, pp 164–175

Gupta RK (2003) The rice–wheat consortium for the Indo-Gangetic Plains: vision and management structure. Addressing resource conservation issues in rice-wheat systems for South Asia: a resource book. RWC-CIMMYT, New Delhi, pp 1–7

Hartemink AE, McBratney A (2008) A soil science renaissance. Geoderma 148:123–129

ICAR-NAAS (Indian Council of Agricultural Research-National Academy of Agricultural Sciences) (2010) Degraded and wastelands of India-Status and spatial distribution. Indian Council of Agricultural Research and National Academy of Agricultural Sciences. Published by the Indian Council of Agricultural Research, New Delhi (56 pp)

Johnson WM (1963) The pedon and polypedon. Soil Sci Soc Am Proc 27:212–215

Kadrekar SB (1979) Utility of basic slag and liming material in lateritic soils of Konkan. Indian J Agron 25:102–104

Kadu PR, Vaidya PH, Balpande SS, Satyavathi PLA, Pal DK (2003) Use of hydraulic conductivity to evaluate the suitability of Vertisols for deep-rooted crops in semi-arid parts of central India. Soil Use Manage 19:208–216

Ming DW, Mumpton FA (1989) Zeolites in soils. In: Dixon JB, Weed SB, Dinauer RC (eds) Minerals in soil environments. Soil Sci Soc Am, Madison, pp 873–911

Nayak DC, Sen TK, Chamuah GS, Sehgal JL (1996a) Nature of soil acidity in some soils of Manipur. J Indian Soc Soil Sci 44:209–214

Nayak DC, Chamuah GS, Maji AK, Sehgal J, Velayutham M (1996b) Soils of Arunachal Pradesh for optimising land use. NBSS Publ. 55b (Soils of India Series), National Bureau of Soil Survey and Land Use Planning, Nagpur, India, 54 pp + one sheet soil map (1:500,000 scale)

NBSS&LUP (1994) Proceedings of national meeting on soil-site suitability criteria for different crops, 7–8 Feb. Nagpur India, p 20

Oates KM, Kamprath EJ (1983) Soil acidity and liming: effects of the extracting solution cations and pH on the removal of aluminium from acid soils. Soil Sci Soc Am J 47:686–690

Pal DK (2012) Pedogenic thresholds in benchmark soils under rice-wheat cropping system in a climosequence of the Indo-Gangetic alluvial plains. Isslup (Indian Society of Soil Survey and Land Use Planning, Nagpur) Special. Publication 2012:35–52

Pal DK (2013) Soil modifiers: their advantages and challenges. Clay Res 32:91–101

Pal DK, Deshpande SB (1987) Characteristics and genesis of minerals in some benchmark Vertisols of India. Pedologie (Ghent) 37:259–275

Pal DK, Dasog GS, Vadivelu S, Ahuja RL, Bhattacharyya T (2000a) Secondary calcium carbonate in soils of arid and semi-arid regions of India. In: Lal R, Kimble JM, Eswaran H, Stewart BA (eds) Global climate change and pedogenic carbonates. Lewis Publishers, Boca Raton, Fl, pp 149–185

Pal DK, Bhattacharyya T, Deshpande SB, Sarma VAK, Velayutham M (2000b) Significance of minerals in soil environment of India. NBSS Review Series, 1. NBSS&LUP, Nagpur, 68 pp

Pal DK, Bhattacharyya T, Ray SK, Bhuse SR (2003a) Developing a model on the formation and resilience of naturally degraded black soils of the peninsular India as a decision support system for better land use planning. NRDMS, Department of Science and Technology (Govt. of India) Project Report, NBSSLUP (ICAR), Nagpur, 144 pp

Pal DK, Srivastava P, Bhattacharyya T (2003b) Clay illuviation in calcareous soils of the semi-arid part of the Indo-Gangetic Plains, India. Geoderma 115:177–192

Pal DK, Bhattacharyya T, Ray SK, Chandran P, Srivastava P, Durge SL, Bhuse SR (2006) Significance of soil modifiers (Ca-zeolites and gypsum) in naturally degraded Vertisols of the peninsular India in redefining the sodic soils. Geoderma 136:210–228

Pal DK, Mandal DK, Bhattacharyya T, Mandal C, Sarkar D (2009a) Revisiting the agro-ecological zones for crop evaluation. Indian J Genet 69:315–318

Pal DK, Bhattacharyya T, Chandran P, Ray SK (2009b) Tectonics-climate-linked natural soil degradation and its impact in rainfed agriculture: Indian experience. In: Wani SP, Rockström J,

Oweis T (eds) Rainfed agriculture: unlocking the potential. CABI International, Oxfordshire, U.K., pp 54–72

Pal DK, Bhattacharyya T, Chandran P, Ray SK, Satyavathi PLA, Durge SL, Raja P, Maurya UK (2009c) Vertisols (cracking clay soils) in a climosequence of Peninsular India: evidence for Holocene climate changes. Quatern Int 209:6–21

Pal DK, Bhattacharyya T, Srivastava P, Chandran P, Ray SK (2009d) Soils of the Indo-Gangetic Plains: their historical perspective and management. Curr Sci 9:1193–1201

Pal DK, Lal S, Bhattacharyya T, Chandran P, Ray SK, Satyavathi PLA, Raja P, Maurya UK, Durge SL, Kamble GK (2010) Pedogenic thresholds in benchmark soils under rice-wheat cropping system in a climosequence of the Indo-Gangetic alluvial plains. Final project report, Division of Soil Resource Studies. NBSS & LUP, Nagpur (193 pp)

Pal DK, Wani SP, Sahrawat KL (2012a) Vertisols of tropical Indian environments: pedology and edaphology. Geoderma 189–190:28–49

Pal DK, Wani SP, Sahrawat KL (2012b) Role of calcium carbonate minerals in improving sustainability of degraded cracking clay soils (Sodic Haplusterts) by improved management: an appraisal of results from the semi-arid zones of India. Clay Res 31:94–108

Pal DK, Bhattacharyya T, Wani SP (2012c) Formation and management of cracking clay soils (Vertisols) to enhance crop productivity: Indian experience. In: Stewart BA (ed) Lal R. World soil resources, Francis and Taylor, pp 317–343

Pal DK, Wani SP, Sahrawat KL (2013a) Zeolitic soils of the Deccan basalt areas in India: their pedology and edaphology. Curr Sci 105:309–318

Pal DK, Sarkar D, Bhattacharyya T, Datta SC, Chandran P, Ray SK (2013b) Impact of climate change in soils of semi-arid tropics (SAT). In: Bhattacharyya T, Pal DK Sarkar D, Wani SP (eds) Climate change and agriculture. Studium Press, New Delhi, pp 113–121

Pal DK, Wani SP, Sahrawat KL, Srivastava P (2014) Red ferruginous soils of tropical Indian environments: A review of the pedogenic processes and its implications for edaphology. Catena 121:260–278. doi:10.1016/j.catena2014.05.023

Pal DK, Wani SP, Sahrawat KL (2015) Carbon sequestration in Indian soils: present status and the potential. Proc Natl Acad Sci Biol Sci (NASB) India 85:337–358. doi:10.1007/s40011-014-0351-6

Rajaguru SN, Korisetter R (1987) Quaternary geomorphic environment and culture succession in western India. Indian J Earth Sci 14:349–361

Randhawa MS (1945) Progressive desiccation of northern India in historical times. J Bombay Nat Hist Soc 45:558–565

Ritter DF (1996) Is quaternary geology ready for the future? Geomorphology 16:273–276

Robinson GH (1971) Exchangeable sodium and yields of cotton on certain clay soils of Sudan. J Soil Sci 22:328–335

Sabale AB, Vishwakarma LL (1996) Zeolites and associated secondary minerals in Deccan volcanics: study of their distribution, genesis and economic importance. National symposium on Deccan Flood Basalts, India, Gondwana. Geol Mag 2:511–518

Sahrawat KL (2003) Importance of inorganic carbon in sequestering carbon in soils of dry regions. Curr Sci 84:864–865

Sahrawat KL (2004) Organic matter accumulation in submerged soils. Adv Agron 81:169–201

Sahrawat KL, Bhattacharyya T, Wani SP, Chandran P, Ray SK, Pal DK, Padmaja KV (2005) Long-term lowland rice and arable cropping effects on carbon and nitrogen status of some semi-arid tropical soils. Curr Sci 89:2159–2163

Sanchez PA (1976) Properties and management of soils in the tropics. Wiley, New York

Sanchez PA, Logan TJ (1992) Myths and science about the chemistry and fertility of soils in the tropics. In: Lal R, Sanchez PA (eds) Myths and science of soils of the tropics SSSA Special Publication Number 29. SSSA, Inc and ACA, Inc, Madison, Wisconsin, USA, pp 35–46

Sen TK, Dubey PN, Chamuah GS, Sehgal JL (1997) Landscape–soil relationship on a transect in central Assam. J Indian Soc Soil Sci 45:136–141

Srivastava P, Pal DK, Aruche KM, Wani SP, Sahrawat KL (2015) Soils of the Indo-Gangetic Plains: a pedogenic response to landscape stability, climatic variability and anthropogenic activity during the Holocene. Earth Sci Rev 140:54–71. doi:10.1016/j.earscirev.2014.10.010

Swindale LD (1989) Approaches to agrotechnology transfer, particularly among Vertisols. In: Management of Vertisols for improved agricultural production: proceedings of an IBSRAM inaugural workshop, 18–22 Feb 1985. ICRISAT Centre, Patancheru, India

Sys C, van Ranst E, Debaeveye J, Beernaert F (1993) Land evaluation, part 3: crop requirements. International Training Centre for Post Graduate Soil Scientists, University of Ghent, Belgium

Tiwary P, Patil NG, Bhattacharyya T, Chandran P, Ray SK, Karthikeyan K, Sarkar D, Pal DK et al (2014) Pedotransfer functions: a tool for estimating hydraulic properties of two major soil types of India. Curr Sci 107:1431–1439

Tomar SS, Tembe GP, Sharma SK, Tomar VS (1996) Studies on some land management practices for increasing agricultural production in Vertisols of central India. Agric Water Manage 30: 91–106

Vaidya PH, Pal DK (2002) Micro topography as a factor in the degradation of Vertisols in central India. Land Degrad Dev 13:429–445

Wani SP, Pathak P, Jangawad LS, Eswaran H, Singh P (2003) Improved management of Vertisols in the semi-arid tropics for increased productivity and soil carbon sequestration. Soil Use Manage 19:217–222

William DW, Cook T, Lynn W, Eswaran H (1996) Evaluating the field morphology of Vertisols. Soil Surv Horiz 37:123–131

Chapter 10
Summary and Concluding Remarks

Abstract The treatise of Indian and tropical soils ends with a chapter 'Summary and Concluding Remarks', which projects a concise but a precious synthesis of unique research results obtained by the soil and earth scientists on major soil types of tropical Indian environments. In the past, much valuable work has been done throughout the tropics, but it has been always difficult to manage these soils to sustain their productivity and it is more so when comprehensive knowledge on their formation remained incomplete for a long time. Soil care continues to be the main issue in national development and thus needs to be a constant research agenda in the Indian context. This is imperative since soil knowledge base becomes critical in meeting the food demand for ever increasing human population. In this task basic pedological research is required to understand some of the unresolved edapho-logical aspects of the tropical Indian soils to develop improved management practices. This chapter highlights the major theme areas of soils (Chaps. 2–9) that have been dealt in the perspective of the recent developments in pedology, min-eralogy, taxonomy and edaphology with context of tectonics and climate change in the Indian sub-continent. The usefulness of such information in unravelling many interesting pedological, edaphological, mineralogical and taxonomical issues of soils of the country has been well established. The synthesis of research results finally transforms to state-of-art information, which may serve as guiding principles to improve and maintain soil health through adequate national recommended practices in other tropical parts of the world.

Keywords Indian tropical soils · State of art information

This treatise provides state-of-art information on recent developments in the tropical soils of India. Summary and concluding remarks of each chapter are highlighted in the following.

1. Under both irrigation and rain-fed conditions Vertisols are cultivated for various agricultural crops. In HT climatic environment, agronomic practices for growing crops under irrigation do not cause soil degradation while in SAD and AD climates crops fail but grow well in the presence of soil modifiers. Without soil

© Springer International Publishing AG 2017
D.K. Pal, *A Treatise of Indian and Tropical Soils*,
DOI 10.1007/978-3-319-49439-5_10

modifiers, however, the soils become saline and sodic under irrigation in SAT environments and lose their productivity. On the other hand, in presence of a soil modifier like palygorskite, non-sodic Vertisols (Typic Haplusterts) have severe drainage problems, like the non-zeolitic Aridic Haplusterts, even with an ESP ≥ 5 but <15. Zeolitic Sodic Haplusterts have no drainage problem and are productive like Typic Haplusterts at present. The present agricultural land uses clearly underscores that even though Vertisols are relatively homogeneous major soil group; they exhibit a remarkable variability in their land use and crop productivity. This scenario calls for the attention of the edaphologists to understand the pedogenetic factors that cause the variability in their properties. A synthesis of recent developments in the pedology of Vertisols achieved through the use of high resolution micro-morphology, mineralogy, and age control data along with their geomorphic and climatic history, has created a much better understanding on the effects of pedogenetic processes due to climate change during the Holocene. The climate change has caused modifications in the soil properties in the presence or absence of Ca-zeolites, gypsum, $CaCO_3$ and palygorskite minerals.

The formation and persistence of Vertisols in the Deccan basalt areas under HT climatic conditions, provides a unique example of tropical soil formation. Such soil formation remained incomprehensible unless the role of zeolites was highlighted by the Indian soil scientists during the last two decades. Persistence of these soils in HT climate for millions of years has provided a deductive check on the inductive reasoning of the conceptual models on the formation of Vertisols in HT climate. Zeolitic Vertisols (both sodic and non-sodic by definition) are under cultivation for crops like rice and sugarcane because of the lack of prolonged waterlogging. Additionally, such soils also support winter crops. These soils under the present typical agricultural land uses are mitigating the adverse effect of Holocene climate change to aridity and also sequestering carbon from the atmosphere. Experimental results obtained on the use of zeolites (other than heulandites) as soil conditioners and slow-release fertilizers provide important clues to address the possible role of soil heulandite in minimizing the conversion NH_4^+ ions to gaseous phases of N and adsorption and desorption of major nutrients in natural soil environments. Delineation of Ca- rich zeolites in Indian soils is now available. Therefore, fresh research efforts are needed to understand the selective role of zeolites in the adsorption and desorption reactions of N, P and K. Such additional knowledge highlights an organic link between pedogenetic processes and bulk soil properties, thereby providing a better understanding of many pedological and edaphological issues related to Vertisols. The novel insights will serve as guiding principles to improve and maintain their health and quality while developing suitable management practices to enhance and sustain their productivity. As a matter of fact, much of the success of the management interventions still depends on the proper classification of Vertisols at the subgroup level, identifying the impairment of drainage in Aridic Haplusterts (ESP ≥ 5, <15), Typic Haplusterts (with palygorskite) and the improvement of drainage in Sodic Haplusterts/Sodic Calciusterts with soil modifiers. The SAT Vertisols at present are less intensively cultivated because of their inherent limitations, despite that they represent a productive resource under improved management. It follows then that geographical areas dominated by Vertisols require immediate national attention for their judicious use to produce more food required for the populous Indian subcontinent and other countries in the developing world.

2. Red ferruginous (RF) soils of tropical environments belong to five taxonomic soil orders (Entisols, Inceptisols, Alfisols, Mollisols and Ultisols). This fact amply justifies a statement that tropical RF soils in India have captured wide soil diversity. The spatially associated Ultisols with acidic Alfisols and Mollisols in both zeolitic and non-zeolitic parent materials in humid tropical climatic environments provides a unique example of tropical soil formation by discounting the exiting conceptual models on tropical soils. However, this fact was not much appreciated, until the role of zeolites and other base rich parent materials was implicated in pedology and edaphology by the Indian researchers during the last two decades. In reality, these soils support multiple production systems and generally maintain positive organic carbon balance without adding significantly to greenhouse gas emissions. A synthesis of literature on the recent developments on the pedology of RF soils, including their physical, chemical, biological, mineralogical and micro-morphological properties, and their degradation status is very timely as the renaissance in soil science is already in place. The new knowledge improves the understanding as to how the parent material composition influences the formation of Alfisols, Mollisols and Ultisols in weathering environments of HT climate. This knowledge also explains how the relict Alfisols of SAT areas is polygenetic created by climate shift during the Holocene. Despite the fact that the extent of soil loss by erosion, and acidity in Ultisols, Alfisols, Mollisols and Inceptisols (with clay enriched B horizons) is generally moderate, these soils need improved nutrient, water and soil water management practices under conservation based agriculture to sustain crop productivity at an enhanced level. Pioneering research efforts have helped establish an inherent link between pedogenetic processes and bulk soil properties, and have facilitated a better comprehension of many pedological and edaphological issues related to Alfisols, Mollisols and Ultisols mainly of HT climate. The synthesis has improved the basic understanding of why the formation of Oxisols from Ultisols is an improbable genetic pathway in tropical environment of India and elsewhere in the world. There is a strong need to modify the mineralogy class of highly weathered RF soils. This treatise will help to dispel some of the myths on the formation of tropical soils and their low fertility by putting in context their characteristics and capacity to be productive. To sustain crop productivity at an enhanced level, large tracts of lands dominated by RF soils need to be brought under improved soil, water and nutrient management to help meet the food needs of ever increasing Indian population.

3. Recent research by both earth scientists and soil scientists in the IGP soils based on large number of well-presented pedons spread along the west hot arid climate

to per-humid climate in the east, have led to new perspectives on the historical development of the IGP and the soils therein. This adequately addresses the hitherto little known subtleties of pedogenesis and polygenesis due to recorded tectonic, climatic and geomorphic episodes and phenomena, and anthropogenic activities during the Holocene.

Based on degree of development, five geomorphic surfaces, QIG1 to QIG5 with soil ages 0.5 ka, 0.5–2.5 ka, 2.5–5.0 ka, 5.0–10 ka, >10 ka respectively, are mappable in the IGP and correspond to the post-incisive chronosequences that evolved in response to interplay of fluvial processes, climatic fluctuations, and neotectonics during the Holocene. The polygenetic signatures, illuvial clay pedofeatures, pedogenic carbonates, clay mineralogy, and stable isotope geochemistry, suggest the evolution of the IGP soils witnessed two humid phases (13.5–11.0 and 6.5–4.0 ka) with intervening dry climatic conditions. The pedogenic response to the neotectonics suggests upliftment of blocks caused break in the sedimentation and initiation of pedogenic activity under the prevailing climate. Episodic uplift of different blocks resulted in a sequence of soils with varying degree of development.

The IGP soils across the topographic gradient (<0.02%) with varying climate from hot-arid to per humid belong to Entisols, Inceptisols, Alfisols, and Vertisols orders. Some of the sodic soils (Natrustalfs) changed to non-sodic soils (Haplustalfs), and the Mollisols (OC enriched soils) changed to Typic Haplustalfs (less OC enriched soils) after two decades of reclamation and agriculture, respectively. Addition and depletion of OC, formation pedogenic $CaCO_3$, illuviation of clay particles and argilli-pedoturbation are the major pedogenic processes in soils of the IGP during the Holocene. The IGP soils are, in general, micaceous, but the soils with vertic characters are smectitic. The soils with micaceous and smectitic mineralogy were formed in alluviums derived from the Himalayas and Cratonic rocks, respectively.

The beginning of agricultural activity over the southwest Asia is represented by the site of Mehrgarh at 7000 BC with further dispersal eastward over the upper and middle Gangetic Plains occurred around 2000 BC. Deforestation and cultivation of the IGP for over several millenniums has influenced the regressive pedogenesis of the IGP. The rapid development of calcareousness and concomitant sub-soil sodicity in semi-arid areas as natural soil degradation process, and enhancement of the $CaCO_3$ (SIC, soil inorganic carbon) and soil bulk density due to anthropogenic activities are the two potential threats, which require appropriate management interventions for restoring and maintaining soil health for sustainable agricultural production.

Soil carbon dynamics can help in determining the pertinence of management interventions of the National Agricultural Research Systems (NARS) to raise as well as to maintain agricultural productivity of soils of the IGP. The sequestration of atmospheric CO_2 as SOC (especially in soils under rice cultivation) and SIC in the vast arid and semi-arid soils suggests that the greenhouse gas emission from the IGP soils do not seem to contribute substantially to the global warming potential.

A better understanding of the pedology of the IGP soils and their linkage to climate change, landscape stability, and anthropogenic activity appear to be potentially useful as guideline for their management. Thus the new knowledge base has potential as a reference for critical assessment of the pedosphere for health and quality in different parts of the world and may facilitate developing a suitable management practices for the food security in the 21st century.

4. The treatise on modelling tropical soils indicates that although among the most popular models applicable in soil formation, the residua and haplosoil models have relevance to formation and persistence of Indian tropical soils, they cannot explain the existence of million years old Vertisols, Alfisols and Mollisols under humid tropical climate because these models did not consider the stability of base rich primary minerals over time. This novel understanding provides a deductive check on the inductive reasoning so far made on the formation of soils in tropical humid climate and also establishes the validity of Jenny's state factor equation in the formation of the Indian tropical soils in the intense weathering environments under HT climate.

5. Pedogenic calcium carbonate, soil sodicity and palygorskite mineral impair the hydraulic properties of the SAT soils, which reduces their crop productivity. This type of unfavourable soil health triggered by the tectonic-climate linked regressive pedogenic processes needs to be globally considered as the natural soil degradation process despite the claims of its occurrence as a result of human induced soil degradation in the SAT areas. The regressive pedogenic processes that are inherently connected to the development of natural soil degradation, expands the basic knowledge in pedology and thus it may have relevance in soils of other SAT areas of the world. Research efforts made in the Indian subcontinent explains the cause-effect relationship of the degradation and provides enough insights as to how the remedial measures are to be invented including the role of pedogenic $CaCO_3$ and geogenic Ca-zeolites as soil modifiers along with gypsum, in making naturally degraded soils resilient and healthy.

6. The mineralogical research work undertaken over the last several decades on important soil/paleosols types and the sediments demonstrates that the pedogenic clay minerals of intermediate weathering stages like HIS, Sm/K, HIV, PCh and pedogenic carbonates can be very useful paleoclimatic indicators. This basic information on soil clay minerals can serve as an important tool for the soil and earth scientists and especially the paleoclimatologists to infer climate change not only of India but elsewhere of the world.

7. A thorough knowledge and appreciation of minerals in soils is critical to our understanding and use of soil. Despite our general understanding on the role of minerals in soils, it is necessary to investigate the properties of the minerals, especially clay minerals, their mixtures and surface modifications in the form that they occur in the soil. From the few examples under different agro-climatic

situations cited it is evident that unless the mineralogical description is accurate enough for the purpose intended, it would not be prudent to look for their significance in soils. With the use of high resolution mineralogy, identification and explanation of many enigmatic situations in soils can be conveniently solved. Therefore, the advanced information developed provides adequate mineralogical database that would explain discretely many unresolved issues of the nutrient management in terms of specific soil minerals in general, and clay minerals in particular and their significance in soil as a sustainable medium for plant growth.

8. Edaphology is inherently based on deep fundamental understandings of soils and thus basic pedological research in tropical soils needs to be encouraged vigorously to link some of their major unresolved edaphological aspects to develop improved management practices.

9. This exposition based on the recent advances in pedology and edaphology of the Indian tropical soils, may serve as guiding principles to improve and maintain soil health through adequate national recommended practices in other tropical parts of the world.